PROBLEM SOLVING IN BIOCHEMISTRY

A Practical Approach

PROBLEM SOLVING IN BIOCHEMISTRY

A Practical Approach

Jane M. Magill

Biochemistry Department
Texas A&M University

Macmillan Publishing Company
NEW YORK

Collier Macmillan Publishing Company
LONDON

Macmillan Publishing Company.
866 Third Avenue, New York, New York 10022

Collier Macmillan Canada, Inc.

ISBN: 0-02-432100-1

Printing: 1 2 3 4 5 6 7 8   Year: 8 9 0 1 2 3 4 5 6 7

This book is dedicated to Clint and Anne

Preface

Problem Solving in Biochemistry is written for both beginning biochemistry students uncomfortable with mathematical applications and for their instructors who must include quantitative reasoning and problem solving as part of the biochemistry course.

Having taught a two-semester biochemistry course for fifteen years, I know that the problems in student comprehension of quantitative concepts come from the disparity in mathematical backgrounds between many of the beginning students and their instructors. Instructors cannot devote lecture time to detailed explanations of elementary mathematics. Nor is there always adequate opportunity to address each student's difficulties in solving assigned problems. *Problem Solving in Biochemistry* is intended to help bridge the gap between the quantitative skills expected by the instructor and the actual skills many students bring to their first biochemistry course.

Problem Solving in Biochemistry consists of five chapters that focus on the presentation and application of fundamental quantitative skills: Tools for Solving Problems in Biochemistry; Acid-Base Chemistry; Enzyme Kinetics; Isolation Purification and Analysis of Proteins and Nucleic Acids; and Bioenergetics. Each chapter begins at an elementary level and proceeds section by section to more complex topics, taking a "how-to-do" approach. Each of these sections includes several example problems whose complete solutions are provided and often discussed. The presentation here recognizes that problem solving requires conceptual knowledge as well as the ability to calculate correctly.

These solved problems are this book's fundamental pedagogical resource. They provide the applied mathematical information and skills, and they tutor the student in the patterns of critical quantitative thinking. As the problems become more complex and challenging in the chapters on enzyme kinetics and bioenergetics, these solutions provide substantial information and explanation to aid the student in interpreting the data and the rationale for the necessary calculations. Step/by/step study of these solved exercises will also build confidence in the students' abililties to solve quantitative problems on their own. Unsolved exercises are provided at the end of each chapter, and the answers for these are given in the back of the book.

In my choice of problems I have included many of comtemporary importance so that students will realize that the skills they are learning are useful and not without practial application to the real world of the practicing biochemist.

Many of the problems in Chapters 3, 4, and 5 apply the knowledge and skills developed in the first two chapters. This encourages the retention of the early material and its application to metabolism, molecular genetics, enzyme kinetics, and bioenergetics.

In each chapter when a new topic is introduced, background information and explanations are provided. However, these are brief, for the focus of *Problem Solving in Biochemistry is on the principles and fundamental applications, not on theory.* This book is designed and written for use together with any of the several comprehensive textbooks of biochemistry necessary for complete mastery of the subject. I have tried to focus this book on the most important topics for an introductory course in biochemistry, and I have emphasized the skills most needed and most lacking.

It is important to realize that this book is not designed to be either a laboratory textbook or a comprehensive treatise on quantitative methods. I hope this book will prove useful in part because it focuses on a few fundamental issues, takes a "how-to-do" approach, provides many solved problems, and does not attempt to do the things for which there are other excellent resources.

At the beginning of this project, I was determined to use only real data in the problems. In most cases this was possible. However, in some problems necessary simplifications have forced me to use realistic rather than real data. Several colleagues at Texas A&M provided me with experimental data which formed the basis for many problems in the book. I am particularily indebted to the following faculty members: Nick Pace for some experimental data and for critical reviews of the chapter on enzyme kinetics; Jim Wild for unpublished data on ATCase; Tom Baldwin for data and discussions on amino acid sequencing; George Bates for discussions on acid-base chemistry; and Clint Magill for suggesting problems in molecular genetics.

This book could not have been completed by this date without the invaluable assistance of Carol Kamps and Lisa Lohman. Carol Kamps, an instructor in biochemistry at Texas A&M, typed the manuscript, checked the problems, and consistenly generated enthusiasm for the project. Lisa Lohman prepared all the illustrations, many of them within 24 hours.

I want to thank the reviewers, particularly James Funston, Tom Laue, and Celia Marshak, whose advice and suggestions improved the book significantly. I also want to thank the staff at Macmillan for their help, especially my editor, Bob Rogers, for his constant encouragement and support from the book's inception.

Jane Magill

Contents

PROBLEM SOLVING IN BIOCHEMISTRY

A Practical Approach

Chapter 1

TOOLS REQUIRED FOR PROBLEM SOLVING IN BIOCHEMISTRY

A large part of any beginning course in biochemistry is the solving of problems. The calculations required for solving these problems demand that the student master several basic concepts in mathematics. Besides these basic math skills, many problems involve units of weight, volume, and concentration commonly used in biochemistry and it is essential that the student become familiar with them. For example, to determine how a protein may behave in the cell, it is helpful to know the size and net charge on the protein under conditions similar to those in the cell. For this, we must do calculations using numbers in exponential form, and determine log and antilog values. To determine the amount of a substance produced by a cell, rates of enzyme catalysis must be used. Studying rates requires setting up ratios and determining concentration changes.

In this first chapter we emphasize the mathematical concepts that are essential to problem solving in biochemistry. The sections on weights, volumes, and concentrations stress only the units most commonly used in biochemistry. The final section, dealing with standard curves, explains the basic concepts in graphing data and in determining concentrations of biochemical compounds from absorbances of known concentrations.

Mathematical Concepts Important in Problem Solving

Most of the difficulties that students encounter in biochemical calculations seem to fall into the following areas: dealing with numbers in exponential form, solving equations, and establishing ratios and proportions.

Exponential Forms

Frequently, it is easier to write a very large number or a very small fraction in *exponential form* with *base* 10. For example, the number 1,000,000 may be written 1.0×10^6, where

1.0 is the coefficient,

10 is the base, and

6 is the exponent or power.

Table 1-1. Common Numbers and Their Corresponding Exponential Forms

$$1,000,000 = 10^6$$
$$100,000 = 10^5$$
$$10,000 = 10^4$$
$$1,000 = 10^3$$
$$100 = 10^2$$
$$10 = 10^1$$
$$1 = 10^0$$

Exponents. *Exponents are powers representing the number of times the base is a factor in the product.* The base is defined as the factor that is multiplied times itself. Another definition frequently used by scientists is: *Exponents are powers representing the number of times the base is multiplied times itself, assuming that* 10 *times itself is* 10^1 *and* 10 *times itself twice is* $10 \times 10 = 10^2$.

In the expression 1.6×10^6, 10 is a factor in the product 6 times (as shown below) and then multiplied by the coefficient, 1.6.

$$1,600,000 = 1.6 \times 10^6 = 1.6 \times 10 \times 10 \times 10 \times 10 \times 10 \times 10$$

Note: For any number greater than 1.0, the exponent (for base 10) must be positive.

Problem 1-1

Express each number in exponential form with base 10.
- (a) 19,600
- (b) 5,352,000
- (c) 11
- (d) 11,300,000

Solution:

(a) $19,600 = 19.6 \times 10^3$ or 1.96×10^4

(b) $5,352,000 = 5.352 \times 10^6$

(c) $11 = 1.1 \times 10^1$

(d) $11,300,000 = 11.3 \times 10^6$ or 1.13×10^7

Problem 1-2

Write each number in exponential form.

- (a) 5×10^1
- (b) 8.1×10^5
- (c) 3.9×10^4
- (d) 0.9×10^6
- (e) 6.02×10^{23} (Avogadro's number)
- (f) 3×10^0

Solution:

(a) $5 \times 10^1 = 50$

(b) $8.1 \times 10^5 = 810,000$

(c) $3.9 \times 10^4 = 39,000$

(d) $0.9 \times 10^6 = 900,000$

(e) $6.02 \times 10^{23} = 602,000,000,000,000,000,000,000$

(f) $3 \times 10^0 = 3 \times 1 = 3$

Note: Any base raised to the zero power is always 1; therefore, $10^0 = 1$.

Fractions may also be expressed in exponential form with base 10. For example, 0.0001 may be written as 1×10^{-4}. Note that *when a number is less than* 1, *the exponential (for base* 10) *becomes a negative value.*

$$
\begin{aligned}
1.0 &= 10^0 \\
0.1 &= 10^{-1} \\
0.01 &= 10^{-2} \\
0.001 &= 10^{-3} \\
0.0001 &= 10^{-4} \\
0.00001 &= 10^{-5} \\
0.000001 &= 10^{-6}
\end{aligned}
$$

Another way of thinking about this is that for values of less than 1, the expression is $1/10^x$, where the exponent x indicates how many times 10 is a factor in the product. Thus

$$0.1 = \frac{1}{10} = \frac{1}{10^1} = 10^{-1}$$

and

$$0.01 = \frac{1}{10^2} \quad \text{or} \quad \frac{1}{10 \times 10} = 10^{-2}$$

Problem 1-3

Express each fraction in exponential form with base 10.

(a) 0.10   (b) 0.12

(c) 0.0038   (d) $\dfrac{1}{1,000,000}$

(e) 0.000031

Solution:

(a) $0.10 = 1.0 \times 10^{-1}$

(b) $0.12 = 1.2 \times 10^{-1}$

(c) $0.0038 = 3.8 \times 10^{-3}$

(d) $\dfrac{1}{1,000,000} = 1.0 \times 10^{-6}$ (e) $0.000031 = 31 \times 10^{-6}$ or 3.1×10^{-5}

Adding and Subtracting Terms in Exponential Form. When adding or subtracting numbers in exponential form, we add or subtract *only the coefficients*, provided that the base and exponent are the same. For example,

$$(1 \times 10^6) + (2 \times 10^6) = (1 + 2) \times 10^6 = 3 \times 10^6$$

If the exponents are different, the numbers may be rewritten so that all have the *same base* and *exponent*. For example,

$$1 \times 10^6 - 2 \times 10^5 = 10 \times 10^5 - 2 \times 10^5$$
$$= (10 - 2) \times 10^5 = 8 \times 10^5$$

Problem 1–4

Add or subtract the following numbers.
  (a) $10^{-5} + 10^{-5}$
  (b) $10^3 + 10^5$
  (c) $5 \times 10^1 + 4.2 \times 10^2$
  (d) $4 \times 10^{-2} - 3 \times 10^{-2}$
  (e) $9.6 \times 10^4 - 3.0 \times 10^3$

Solution:
  (a) $10^{-5} + 10^{-5} = (1 + 1) \times 10^{-5} = 2 \times 10^{-5}$
  (b) $10^3 + 10^5 = (1 \times 10^3) + (100 \times 10^3) = 101 \times 10^3$
      $= 1.01 \times 10^5$
  (c) $5 \times 10^1 + 4.2 \times 10^2 = (5 \times 10^1) + (42 \times 10^1)$
      $= 47 \times 10^1$ or 470
  (d) $4 \times 10^{-2} - 3 \times 10^{-2} = (4 - 3) \times 10^{-2} = 1 \times 10^{-2}$
  (e) $9.6 \times 10^4 - 3.0 \times 10^3 = 96.0 \times 10^3 - 3.0 \times 10^3$
      $= 93.0 \times 10^3$

Multiplying and Dividing Terms in Exponential Form (with Base 10). When numbers in exponential form are multiplied, *their coefficients are multiplied* but *their exponent values are added*. For example,

$$(2 \times 10^6) \times (4 \times 10^3) = (2 \times 4)(10^{6+3}) = 8 \times 10^9$$

When numbers in exponential form are divided, *their coefficients are divided* but *their exponents are subtracted.* For example,

$$\frac{2 \times 10^6}{4 \times 10^3} = \frac{2}{4}(10^{6-3}) = 0.5 \times 10^3$$

Problem 1-5

Multiply or divide as indicated by the sign.

(a) $(7.3 \times 10^6) \times (1.0 \times 10^5)$ (b) $(7.3 \times 10^6) \div (1.0 \times 10^5)$

(c) $(5.1 \times 10^{-3}) \times (9.8 \times 10^{-6})$ (d) $(5.1 \times 10^{-3}) \div (9.8 \times 10^{-6})$

(e) $(4.2 \times 10^{-4}) \times (6.3 \times 10^{-5})$ (f) $(4.2 \times 10^{-4}) \div (6.3 \times 10^{-5})$

Solution:

(a) $(7.3 \times 1.0)(10^{6+5}) = 7.3 \times 10^{11}$ Coefficients are multiplied but exponents are added.

(b) $\frac{7.3}{1.0} \times 10^{6-5} = 7.3 \times 10^1 = 73$ Coefficients are divided but exponents are subtracted.

(c) $(5.1 \times 9.8)(10^{-3+(-6)}) = 49.98 \times 10^{-9}$

(d) $\frac{5.1 \times 10^{-3}}{9.8 \times 10^{-6}} = \frac{5.1}{9.8}(10^{-3-(-6)}) = 0.52(10^{-3+6})$
$$= 0.52 \times 10^3 = 520$$

(e) $(4.2 \times 10^{-4}) \times (6.3 \times 10^{-5}) = (4.2 \times 6.3)(10^{-4+(-5)})$
$$= 26.46(10^{-9})$$

(f) $\frac{4.2 \times 10^{-4}}{6.3 \times 10^{-5}} = \frac{4.2}{6.3}(10^{-4-(-5)}) = 0.67(10^1) = 6.7$

Problem 1-6

Simplify each expression.

(a) $\frac{10^{-4} \times 10^{-5}}{10^{-3}}$ (b) $\frac{10^{-6}}{10^{-3} \times 10^{-5}}$

(c) $\frac{(10^2)(10^{-5})}{10^6}$ (d) $18,900 \times 5,800,000,000$

(e) $\frac{(0.72 \times 10^{-4}) \times (14 \times 10^{-7})}{33.2 \times 10^{-6}}$ (f) $\frac{(9.1 \times 10^{-5}) + (3 \times 10^{-4})}{1.3 \times 10^{-3}}$

(g) $\frac{(4.1 \times 10^{-3}) - (2.9 \times 10^{-4})}{5.6 \times 10^{-5}}$

Solution:

(a) $\dfrac{(10^{-4}) \times (10^{-5})}{10^{-3}} = \dfrac{10^{-9}}{10^{-3}} = 10^{[-9-(-3)]} = 10^{-9+3} = 10^{-6}$

(b) $\dfrac{10^{-6}}{(10^{-3}) \times (10^{-5})} = \dfrac{10^{-6}}{10^{-8}} = 10^{-6-(-8)}$
$= 10^{-6+8} = 10^{2}$

(c) $\dfrac{(10^{2})(10^{-5})}{10^{6}} = \dfrac{10^{-3}}{10^{6}} = 10^{-3-6} = 10^{-9}$

(d) $(1.89 \times 10^{4}) \times (5.8 \times 10^{9}) = 10.96 \times 10^{4+9}$
$= 10.96 \times 10^{13}$ or 1.096×10^{14}

(e) $\dfrac{(0.72 \times 10^{-4})(14 \times 10^{-7})}{33.2 \times 10^{-6}} = \dfrac{10.08 \times 10^{-4+(-7)}}{33.2 \times 10^{-6}}$
$= \dfrac{10.08 \times 10^{-11}}{33.2 \times 10^{-6}} = 0.304 \times 10^{-11-(-6)}$
$= 0.304 \times 10^{-11+6} = 0.304 \times 10^{-5}$

(f) $\dfrac{(9.1 \times 10^{-5}) + (3 \times 10^{-4})}{1.3 \times 10^{-3}} = \dfrac{(9.1 \times 10^{-5}) + (30 \times 10^{-5})}{1.3 \times 10^{-3}}$
$= \dfrac{39.1 \times 10^{-5}}{1.3 \times 10^{-3}}$
$= 30.08 \times 10^{-5-(-3)} = 30.08 \times 10^{-5+3}$
$= 30.08 \times 10^{-2}$

(g) $\dfrac{(4.1 \times 10^{-3}) - (2.9 \times 10^{-4})}{5.6 \times 10^{-5}} = \dfrac{(41 \times 10^{-4}) - (2.9 \times 10^{-4})}{5.6 \times 10^{-5}}$
$= \dfrac{38.1 \times 10^{-4}}{5.6 \times 10^{-5}}$
$= 6.8 \dfrac{10^{-4}}{10^{-5}} = 6.8 \times 10^{-4-(-5)}$
$= 6.8 \times 10^{-4+5} = 6.8 \times 10^{1} = 68$

Logarithms. The *logarithm* (or *log*) is that exponent which indicates the number of times the base is a factor in the product *when the coefficient is* 1.0. For example, the number 2500 may be written in several ways, as shown below using the base 10:

$$2500 = 2.5 \times 10^{3} = 1.0 \times 10^{3.39} = 10^{3.39}$$
$$\quad\quad\quad (1) \quad\quad\quad\quad (2) \quad\quad\quad (3)$$

In version 1, the 3 is an exponent but it is *not* the log. In versions 2 and 3, the exponent, 3.39, is the log since the coefficient is 1.0. By definition, when 10 is a factor in the product 3.39 times, the number is 2500.

The $\log_{10}$ or log symbol indicates logarithms with the base 10. Other bases may be used in biochemical calculations, but 10 is the most common. The base

$$e = 2.718$$

is used less frequently. Logarithms using base e, called natural logarithms, are written ln. Logarithm values may be determined using $\log_{10}$ or ln tables but are generally found using a personal calculator.

Determining $\log_{10}$ Values with a Calculator. There are several different types of personal calculators in use today. The most popular types fall into three categories in terms of the method for obtaining $\log_{10}$ values. For any particular model see the instructions that accompany it.

Type 1.   Most personal calculators are in this category. Press number keys and then the $\boxed{\text{log}}$ key. The $\log_{10}$ of that number will be displayed.

Type 2.   Texas Instruments model 59 is an example. This type requires a second function key. For example, the number keys are pressed and then the $\boxed{\text{2nd}}$ and $\boxed{\text{ln x}}$ keys are pressed, in that order. The $\log_{10}$ of the number entered will be displayed.

Type 3.   Examples include some Hewlett-Packard models. These calculators require the number be entered by pressing ENTER after the number keys, followed by:

(a) the function key (may be color-coded to correspond to $\log_{10}$) and

(b) the $\boxed{\dfrac{10^x}{\log}}$ key

The number displayed is the $\log_{10}$ of the number entered.

Problem 1-7

Determine $\log_{10}$ values for each number without using a calculator.

    (a) 100   (b) 1.0

    (c) 1,000,000   (d) 0.001

    (e) 0.01

Solution:

    (a) $100 = 10^2$ $\log_{10} = 2$

    (b) $1.0 = 10^0$ $\log_{10} = 0$

    (c) $1,000,000 = 10^6$ $\log_{10} = 6$

    (d) $0.001 = 10^{-3}$ $\log_{10} = -3$

    (e) $0.01 = 10^{-2}$ $\log_{10} = -2$

Problem 1-8

Determine $\log_{10}$ values for each number using a calculator.

(a) 123   (b) 2.6

(c) 4.51×10^{-2} (d) 9.07×10^{-4}

(e) 5.56×10^3

Solution:

(a) $\log 123 = 2.09$ (i.e., $123 = 10^{2.09}$)

(b) $\log 2.6 = 0.415$ (i.e., $2.6 = 10^{0.415}$)

(c) $\log (4.51 \times 10^{-2}) = \log 0.0451 = -1.34$

(d) $\log (9.07 \times 10^{-4}) = \log 0.000907 = -3.04$

(e) $\log (5.56 \times 10^3) = \log 5560 = 3.74$

Problem 1-9

Determine $\log_{10}$ for each number from 1 to 10.

Solution:

Number	$\log_{10}$	Number	$\log_{10}$
1	0.000	6	0.778
2	0.301	7	0.845
3	0.497	8	0.903
4	0.602	9	0.954
5	0.699	10	1.000

Problem 1-10

Determine $\log_{10}$ value for each number using a calculator.

(a) 35   (b) 435

(c) 0.957   (d) 4

(e) 0.082   (f) 0.0082

(g) 2.8×10^{-7}

Solution:

(a) $\log 35 = +1.54$

(b) $\log 4.35 = +2.63$

(c) $\log 0.957 = -0.019$

(d) $\log 4 = +0.60$

(e) $\log 0.082 = -1.086$

(f) $\log 0.0082 = -2.086$

(g) $\log (2.8 \times 10^{-7}) = \log 0.00000028 = -6.55$ (or $\log 2.8 + \log 10^{-7} = -6.55$)

Note: Adding log values is the same as multiplying the numbers they represent.

The example below illustrates that when the logs of numbers are added, the resultant log represents the log of the numbers themselves multiplied together.

In the expression $(1 \times 10^{-4})(1 \times 10^{-1})$, -4 is the log of (1×10^{-4}) and -1 is the log of (1×10^{-1}).

Since all logs are exponents, we may add the log values to obtain

$$-4 + (-1) = -5$$

Thus -5 is the log of the expression

$$(1 \times 10^{-4})(1 \times 10^{-1})$$

Problem 1-11

Find the $\log_{10}$ value of each expression.

(a) 3950×450

(b) 161×10^{-5}

(c) 5.24×10^{-8}

Solution:

(a) 3950×450 may be rewritten as $(3.95 \times 10^3)(4.5 \times 10^2)$. Multiplying coefficients and adding exponentials, we obtain

$$17.77 \, (10^{3+2}) = 17.77 \, (10^5)$$
$$\log (17.77 \times 10^5) = \log 17.77 + 5$$
$$= 1.25 + 5$$
$$= 6.25$$

(b) $\log (161 \times 10^{-5}) = \log 161 + (-5)$
$= 2.2 - 5$
$= -2.8$

(c) $\log (5.24 \times 10^{-8}) = \log 5.24 + \log 10^{-8}$
$= \log 5.24 + (-8)$
$= 0.72 - 8$
$= -7.28$

Note: Subtracting log values is the same as dividing the numbers they represent.

For example, in the expression $1 \times 10^{-4} \div 1 \times 10^{-6}$, we divide coefficients and subtract exponents, giving

$$\frac{1}{1}[10^{-4 - (-6)}] = 1 \times 10^{+2}$$

The log of $1 \times 10^{+2}$ is +2.

Since all logs are exponents, we may subtract the log of 1×10^{-6} from 1×10^{-4} and obtain the log of the quotient.

$$-4 - (-6) = +2$$

Therefore, the log of the expression $\dfrac{1 \times 10^{-4}}{1 \times 10^{-6}} = +2$.

Problem 1–12

Find the $\log_{10}$ value of each expression.

(a) $\dfrac{0.34 \times 10^{-5}}{1.18 \times 10^2}$ (b) $\dfrac{4.5 \times 10^6}{3.3 \times 10^7}$

(c) $\dfrac{(0.01)(1.05 \times 10^{-5})}{9.81 \times 10^{-5}}$

Solution:

(a) $\log (0.34 \times 10^{-5}) - \log (1.18 \times 10^2) = \log \dfrac{0.34 \times 10^{-5}}{1.18 \times 10^2}$
$= (\log 0.34 + \log 10^{-5}) - (\log 1.18 + \log 10^2)$
$= [-0.47 + (-5)] - (0.07 + 2)$
$= (-0.47 - 5) - (2.07)$
$= -7.54$

(b) $\log \dfrac{4.5 \times 10^6}{3.3 \times 10^7} = \log(4.5 \times 10^6) - \log(3.3 \times 10^7)$

$= (\log 4.5 + \log 10^6) - (\log 3.3 + \log 10^7)$

$= (0.65 + 6) - (0.52 + 7)$

$= 6.65 - 7.52$

$= -0.87$

(c) $\log \dfrac{(0.01)(1.05 \times 10^{-5})}{9.81 \times 10^{-5}}$

$= \log(10^{-2}) + \log(1.05) + \log(-5) - (\log 9.81 + \log 10^{-5})$

$= [-2 + 0.02 + (-5)] - [0.99 + (-5)]$

$= (-1.98 - 5) - (-4.01)$

$= -6.98 + 4.01$

$= -2.97$

Problem 1-13

Simplify each expression and then find $\log_{10}$.

(a) $\log(5.1 \times 10^{-4})(3.2 \times 10^{-3})$

(b) $\log \dfrac{(4.3 \times 10^{-6})(4.8 \times 10^{-5})}{1.1 \times 10^{-3}}$

(c) $\log \dfrac{(7.7 \times 10^{-2})(8.3 \times 10^{-1})}{(3.0 \times 10^{-1})(5.9 \times 10^2)}$

Solution:

(a) $\log(5.1 \times 10^{-4})(3.2 \times 10^{-3})$

$= \log(5.1 \times 3.2 \times 10^{-4+(-3)})$

$= \log(16.32 \times 10^{-7})$

$= 1.21 + (-7)$

$= -5.79$

(b) $\log \dfrac{(4.3 \times 4.8 \times 10^{-6+(-5)-(-3)})}{1.1}$

$= \log(18.76 \times 10^{-8})$

$= \log 18.76 + \log 10^{-8}$

$= 1.27 + (-8)$

$= -6.73$

(c) $\log \dfrac{(7.7 \times 10^{-2})(8.3 \times 10^{-1})}{(3.0 \times 10^{-1})(5.9 \times 10^2)}$

$= \log \dfrac{(7.7 \times 8.3 \times 10^{[-2+(-1)] - [(-1)+2]})}{3.0 \times 5.9}$

$= \log 3.61 + \log 10^{-4}$

$= 0.56 + (-4)$

$= -3.44$

Determining Natural Logarithms (ln) with a Calculator. Natural logs (ln) use the base $e = 2.718$ instead of base 10. The ln of any number may be determined using a personal calculator with this function. Almost all calculators require that the number be entered and the $\boxed{\text{ln}}$ or $\boxed{\text{ln } x}$ key be pressed. The number displayed is the ln (base e) of the number entered.

Problem 1-14

Determine the ln value for each number from 1 to 10.

Solution:

Number	ln (base e = 2.718)	Number	ln
1	0.000	6	1.791
2	0.693	7	1.946
3	1.098	8	2.079
4	1.386	9	2.197
5	1.609	10	2.303

Converting Natural Logs (ln) to $\log_{10}$. Since the ln of 10 is 2.303 and the $\log_{10}$ of 10 is 1.000, we can write the following ratio:

$$\frac{\ln 10}{\log 10} = \frac{2.303}{1}$$

Thus $\ln 10 = 2.303 \times \log 10$.

Problem 1-15

Convert each natural log to a $\log_{10}$ value. Compare the value of each expression determined as ln and $\log_{10}$.

(a) ln 38

(b) ln (1.9)(298)(0.18)

Solution:

(a) ln 38 = 3.63

ln 38 = 2.303 $\log_{10}(38)$
2.303 $\log_{10}(38)$ = 2.303(1.58) = 3.63

(b) ln (1.9)(298)(0.18) = ln (101.9) = 4.6

ln (1.9)(298)(0.18) = 2.303 $\log_{10}$(1.9)(298)(0.18)

$$2.303 \log_{10}(1.9)(298)(0.18) = 2.303 \log_{10}(101.9)$$
$$2.303(2.0) = 4.6$$

Antilogarithms.   An *antilogarithm* or *antilog* is the number that the log values represents. In one sense, antilogs are the reverse of logs (see Table 1-2).   For example, if we use the number 2500, it may be written as

2.500×10^3 (exponential form)

$10^{3.39}$ (log form), where 3.39 is the $\log_{10}$ of the number 2500

The antilog of 3.39 is 2500.

Table 1-2. Antilogs Corresponding to Log Values[a]

Antilog	$\log_{10}$
10	1
100	2
1000	3
10,000	4
0.1	−1
0.01	−2
0.001	−3
0.0001	−4

[a] Antilogs of positive log values will be numbers greater than 1.0. Antilogs of negative log values will be fractions less than 1.

Determining Antilogs Using a Personal Calculator

Type 1.   Press the number keys and the INV, followed by the log key or 10^x key.

Type 2.   Press number keys, then INV, 2nd, and ln x in that order.

Type 3.   Press number keys, then ENTER, and then the 10^x or log key.

Note: On most calculators, a negative log number is entered as a positive value and the sign is changed using the +/− Key. Then the antilog of the negative value is obtained by the same method as for positive log values.

Problem 1-16

Find the antilog represented by each log value.

(a) 1.4

(b) $-5 + (-3) - (-4)$

(c) 0.08

Solution:

(a) antilog (1.4) = 25.1 (i.e., $10^{1.4}$ = 25.1)

(b) antilog $[-5 + (-3) - (-4)]$ = antilog $(-5 - 3 + 4)$ = antilog (-4) = 0.0001 (i.e., 10^{-4} = 0.0001)

(c) antilog (0.08) = 1.20 (i.e., $10^{0.08}$ = 1.20)

Problem 1-17

Find the antilog of each log value.

(a) -1.0

(b) -0.27

(c) -2.99

(d) -0.6

(e) -3.54

Solution:

(a) antilog $(-1.0) = 1 \times 10^{-1} = 0.1$

(b) antilog $(-0.27) = 0.537$

(c) antilog $(-2.99) = 1.02 \times 10^{-3}$ (or 0.00102)

(d) antilog $(-0.6) = 0.251$

(e) antilog $(-3.54) = 2.88 \times 10^{-4}$ (or 0.000288)

Solving Equations with One Unknown (or Variable)

An equation is composed of terms, some of which are unknowns or variables. For example, in the equation

$$12x + 4 = 0$$

$12x$ is a variable term; 4 and 0 are terms that are constant.

The goal is solving an equation is to isolate the variable (or unknown) so that its coefficient is 1.0.

Coefficients of variables. The coefficient is the number part of the variable term. In the expression $12x$, 12 is the coefficient of the variable, x.

Note: When the coefficient of a variable is 1.0, that variable is written with no coefficient. In the expression

$$12x + x = 2$$

x alone means $1.0x$. Then the expression simplifies to $13x = 2$.

Adding and Subtracting Terms. Terms that are alike may be combined by adding or subtracting their numerical coefficients *only* if the variable portions of the terms are identical. For example, in simplifying the expression

$$4x - 2x + 18x = -7$$

we can subtract $2x$ from $4x$ and then add $18x$ because the variable is x in each term. Thus the expression becomes $20x = -7$.

Transposing Terms. When the variable terms are on opposite sides of the equation, we must transpose in order to isolate the variable. *In transposing terms, it is important to perform identical operations to both sides of the equation.*
For example, in simplifying the expression

$$4x - 2x = x + 7$$

we must combine like variables first to give

$$2x = x + 7$$

In this case, we must transpose x from the right side to the left side to isolate the variable. The same operation is performed on both sides of the equation.

$$2x - (x) = x - (x) + 7 \qquad (x \text{ subtracted from both sides})$$
$$x = 0 + 7$$
$$x = +7$$

Since a value subtracted from itself is always zero, we omit writing this and simply show the value subtracted from the other side.

$$2x = x + 7$$
$$2x - x = +7$$
$$x = +7$$

Problem 1-18

Simplify the expression $19x + 8 + x = 150 + 10$.

Solution: Rewriting gives us

$$19x + x + 8 = 160$$
$$20x = 160 - 8$$
$$20x = 152$$

Similarly, to transpose a negative value, we add that value to each side of the equation. In simplifying the expression

$$46x + 1.5x - 103 = 0$$

we collect like terms:

$$47.5x - 103 = 0$$

Then we add 103 to each side:

$$47.5x = 0 + 103$$
$$47.5x = 103$$

Multiplying and Dividing Terms. When solving for the value of a variable, we try to rearrange the expression so that the coefficient of that variable is 1. In performing any operation with an equation, it is important to remember that both sides of the equation must be treated in an identical way.

In solving the following equation for x, we first combine like terms and then divide each side by the coefficient of x.

$$4x + 18 = 16x - 5$$
$$4x - 16x = -18 - 5$$
$$-12x = -23$$
$$23 = 12x$$

Dividing each side of the equation by the coefficient of x, we obtain

$$\frac{12x}{12} = \frac{23}{12} \quad \text{or} \quad x = \frac{23}{12} \quad \text{then } x = 1.916$$

Since dividing the coefficient of any term by itself always gives 1, we omit writing this step and show only the division of the other side of the equation.

$$x = \frac{23}{12}; \quad x = 1.916$$

Problem 1-19

Solve each equation for x.

(a) $118x - 10 = x$

(b) $\dfrac{0.83}{x} = \dfrac{0.14}{0.002}$

(c) $-1.5 = -\dfrac{1}{2.5\,(1 + \dfrac{60}{x})}$

Solution:

(a) $118x - x = 10$

$\qquad 117x = 10$

$\qquad\quad x = \dfrac{10}{117}$

$\qquad\qquad = 0.085$

(b) $\dfrac{0.83}{x} = \dfrac{0.14}{0.002}$

$\quad 0.83(0.002) = 0.14x$

$\qquad x = \dfrac{0.00166}{0.14}$

$\qquad\quad = 0.0118$

(c) $-1.5 = -\dfrac{1}{2.5\,(1 + \dfrac{60}{x})}$

Multiply both sides by 2.5:

$$2.5(1.5) = \dfrac{1}{1 + 60/x}$$

$$3.75 + \dfrac{225}{x} = 1$$

$$\dfrac{225}{x} = -2.75$$

$$x = -\dfrac{225}{2.75} = -81.8$$

Problem 1-20

In each problem, solve for x.

(a) $4.5 = 4.1 + \log \dfrac{x}{25 - x}$

(b) $6.3 = 7.1 + \log \dfrac{50 - x}{x}$

(c) $0 = +1.9 + 1.363 \log \dfrac{(2 \times 10^{-4})(1.1 \times 10^{-3})}{10^{-x}}$

Solution:

(a) $4.5 = 4.1 + \log \dfrac{x}{25 - x}$

$0.4 = \log \dfrac{x}{25 - x}$ $\text{antilog}\,(0.4) = \dfrac{x}{25 - x}$

$\text{antilog}\,(0.4) = 2.5$

$2.5 = \dfrac{x}{25 - x}\,;\, 62.5 - 2.5x = x$

$62.5 = 3.5x$

$x = \dfrac{62.5}{3.5} = 17.86$

(b) $6.3 = 7.1 + \log \dfrac{50 - x}{x}$

$-0.8 = \log \dfrac{50 - x}{x}$; $\text{antilog}\,(-0.8) = \dfrac{50 - x}{x}$

$\text{antilog}\,(-0.8) = 0.16$

$0.16 = \dfrac{50 - x}{x}$ $0.16x = 50 - x$

$1.16x = 50$

$x = \dfrac{50}{1.16} = 43.1$

(c) $0 = +1.9 + 1.363 \log \dfrac{(2 \times 10^{-4})(1.1 \times 10^{-3})}{10^{-x}}$

$-1.9 = 1.363 \log (2 \times 1.1)(10^{-4}) + (-3) - (-x)$

$\dfrac{-1.9}{1.363} = \log (2.2) + \log (10^{-7} + x)$

Taking log values,

$-1.9 = 1.363 \left(\dfrac{-3.69 + (-2.96)}{-x} \right)$

$\dfrac{-1.9(-x)}{1.363} = -3.69 - 2.96$

$-1.393x = -6.65$

$x = 6.65/1.393$

$\quad = 4.77$

Ratio and Proportion

Many biochemical calculations involve setting up ratios and then establishing correct proportions.

1. A *ratio* is a fraction that allows comparison of two quantities. All concentrations are ratios of amount of solute to amount of solvent or solution. All rates of chemical reactions are ratios of amount of product formed per time unit (e.g., micromoles of product per minute).

2. A *proportion* is an equation which states that two ratios are equal; for example, a proportion may be set up to determine the amount of solute needed to form a solution of the desired concentration.

If 100 mL of a 4.8 M solution of aspartic acid in water is desired, the following proportion or equation may be set up to determine how much aspartic acid to add to 100 mL (or 0.1 L):

$$4.8\ M = \frac{4.8\ \text{mol}}{1.0\ \text{L}} = \frac{x\ \text{moles}}{0.100\ \text{L}}$$

This equation has one unknown (or variable), x, and by solving the equation we find that $x = 0.48$ mol of aspartic acid.

Problem 1-21

Express as a ratio and simplify to give the concentration.

(a) 0.08 mol of glucose in 100 mL of solution

(b) 1.92 mol of glucose in 1.1 L of solution

(c) 70 g of glucose in 1 L of solvent

Solution:

(a) $\dfrac{0.08\ \text{mol glucose}}{100\ \text{mL solution}} = \dfrac{0.08\ \text{mol}}{0.10\ \text{L}} = \dfrac{0.8\ \text{mol}}{1\ \text{L}} = 0.8\ M$

(b) $\dfrac{1.92\ \text{mol glucose}}{1.1\ \text{L solution}} = \dfrac{1.74\ \text{mol}}{1\ \text{L}} = 1.74\ M$

(c) $\dfrac{70\ \text{g glucose}}{1\ \text{L solvent}} = \dfrac{7.0\ \text{g}}{100\ \text{mL}} = 7\%\ \text{glucose}$

Problem 1-22

If 0.01 mg of an enzyme causes 5×10^{-6} mol of glucose to be converted to glucose-6-phosphate (G-6-P) per minute, how much glucose would be converted by 1.5×10^{-4} g of enzyme in 1 min at the same rate?

Solution:

Let the amount of glucose to be converted equal x. Since the *rates are the same,* the proportion can be written

$$\frac{(5 \times 10^{-6})/1.0\ \text{min}}{10^{-5}\ \text{g enzyme}} = \frac{x/1.0\ \text{min}}{1.5 \times 10^{-4}\ \text{g enzyme}}$$

or

$$\frac{5 \times 10^{-6}}{10^{-5} \text{ g enzyme}} = \frac{x}{1.5 \times 10^{-4} \text{ g enzyme}}$$

Simplifying yields

$$(5 \times 10^{-6})(1.5 \times 10^{-4}) = 10^{-5} \cdot x$$

$$\frac{(5 \times 1.5)(10^{-6+(-4)})}{10^{-5}} = x$$

$$x = \frac{7.5 \times 10^{-10}}{10^{-5}} = 7.5 \times 10^{(-10+5)}$$

$$= 7.5 \times 10^{-5}$$

Thus 7.5×10^{-5} mol of glucose would be converted to G-6-P by 1.5×10^{-4} g of enzyme.

Units of Measurement Used in Biochemical Calculations

Units of Mass and Weight

Mass is a measure of the amount of a substance. The *weight* of a substance is a measure of its mass and varies with earth's gravity. Weights are generally used as a measure of mass in biochemical calculations.

The *gram* is the basic unit of weight measurement in the sciences. However, in biochemistry, fractions of grams are most often required, so that the following units are more common.

$$1 \text{ milligram (mg)} = \frac{1}{1000} \text{ g or } 0.001 \text{ g}$$

$$1 \text{ microgram } (\mu\text{g}) = \frac{1}{1{,}000{,}000} \text{ g or } 0.000001 \text{ g}$$

$$1 \text{ nanogram (ng)} = \frac{1}{1{,}000{,}000{,}000} \text{ g or } 0.000000001 \text{ g}$$

It is more convenient to express these fractions in exponential form with base 10.

$$1.0 \text{ g} \quad = 10^0 \text{ g}$$

$$1.0 \text{ mg} = 10^{-3} \text{ g}$$

$$1.0 \ \mu\text{g} \ = 10^{-6} \text{ g}$$

$$1.0 \text{ ng} \ = 10^{-9} \text{ g}$$

Problem 1-23

Show how many g, mg, and ng are represented by 4515 μg.

Solution

$$4515 \ \mu g = 0.004515 \ g \text{ or } 4.515 \ mg \text{ or } 4,515,000 \ ng \ (4.515 \times 10^6 \ ng)$$

or

$$4515 \ \mu g = 4515 \times 10^{-6} \ g = 4.515 \times 10^{-3} \ g$$

Problem 1-24

Express each gram quantity in exponential form with base 10.

  (a) 0.01 g

  (b) 1.0 μg

  (c) 10 g

  (d) 0.0001 g

  (e) 0.002 g

Solution:

  (a) 0.01 g = 10^{-2} g

  (b) 1.0 μg = 10^{-6} g

  (c) 10 g = 10^{+1} g

  (d) 0.0001 g = 1×10^{-4} g

  (e) 0.002 g = 2×10^{-3} g

A *mole* is defined as 6.02×10^{23} molecules of a pure substance. The *molecular weight of a pure substance* is the *weight* in grams of 6.02×10^{23} molecules of that substance.

Biochemical compounds are composed mainly of the following elements: C, H, O, N, P, S, Fe, Ca^{2+}, Mg^{2+}. To determine the weight of 1.0 mol of a compound, the sum of the atomic weights of all atoms in the molecule must be determined. The atomic weights are some elements are listed in Table 1-3.

Table 1-3. Atomic Weights of Elements Most Commonly Found in Biochemical Compounds

Element	Atomic Symbol	Atomic Weight
Hydrogen	H	1
Carbon	C	12
Oxygen	O	16
Nitrogen	N	14
Phosphorus	P	31
Sulfur	S	32
Iron	Fe	56
Calcium	Ca	40

Problem 1-25

Calculate from Table 1-3 the weight of 1.0 mol of the amino acid alanine:

$$CH_3-CH-COO^-$$
$$\quad\; |$$
$$\quad\; {}^+NH_3$$

Solution: The atomic weights of each atom in alanine may be found from Table 1-3.

Atom	Atomic Weight	Total in Molecule	Total Weight
C	12	3	36
H	1	7	7
O	16	2	32
N	14	1	14
		Sum =	89

The weight of 1.0 mol of alanine is 89 g.

Problem 1-26

How many molecules are in 0.12 mol of alanine?

Solution:

Let the number of molecules be x. Since each mole *by definition* contains 6.02×10^{23} molecules, 0.12 mol of alanine contains

$$\frac{0.12 \text{ mol}}{1.0 \text{ mol}} = \frac{x}{6.02 \times 10^{23}}$$

Thus

$$x = \frac{0.12(6.02 \times 10^{23})}{1} = 0.722 \times 10^{23} \text{ molecules}$$

Problem 1-27

How many molecules are in 5 ng of pure alanine?

Solution:

Let x = the number of molecules in 5 ng. 5 ng represents $(5 \times 10^{-9})/89$ of the weight of 1.0 mol of alanine. Since 1.0 mol contains 6.02×10^{23} molecules, the proportion may be written

$$\frac{1.0}{6.02 \times 10^{23}} = \frac{(5 \times 10^{-9})/89}{x}$$
$$x = \frac{(6.02 \times 10^{23})(5 \times 10^{-9})}{89}$$
$$= 0.338 \times 10^{14} = 3.38 \times 10^{13} \text{ molecules}$$

Fractions of a Mole

In a cell, compounds are present in fractions of mole quantities. Generally, only 10^{-12} mol (picomole) or at most 10^{-9} mol (nanomole) of any substance is found in a cell. In most biochemical reactions measured in test tubes, small amounts of cell extracts or cellular compounds are used. The fractions of a mole most commonly used in biochemical calculations are listed in Table 1-4.

Table 1-4. Fractions of a Mole

Fraction of Mole	Common Name	Abbreviation
0.001 or 10^{-3}	millimole	mmol
0.000001 or 10^{-6}	micromole	μmol
0.000000001 or 10^{-9}	nanomole	nmol
0.000000000001 or 10^{-12}	picomole	pmol

Problem 1-28

Write each amount in moles, expressing values in exponential form.

   (a) 0.3 mmol   (b) 4.5 μmol

   (c) 1.9 nmol   (d) 10.8 pmol

Solution:

   (a) 0.3 mmol = 0.3×10^{-3} mol or 3.0×10^{-4} mol

   (b) 4.5 μmol = 4.5×10^{-6} mol

   (c) 1.9 nmol = 1.9×10^{-9} mol

   (d) 10.8 pmol = 10.8×10^{-12} mol or 1.08×10^{-13} mol

Problem 1-29

What is the weight, in grams, of each of the following amounts of the amino acid glycine (MW = 75)?

(a) 1.4 mmol

(b) 0.03 mol

(c) 22 μmol

(d) 79 nmol

(e) 15 pmol

Solution:

(a) 1.4 mmol weighs $1.4 \times 10^{-3} \times 75$ g = 105×10^{-3} g or 1.05×10^{-2} g

(b) 0.03 mol weighs 0.03×75 g = 2.25 g

(c) 22 μmol weighs $22 \times 10^{-6} \times 75$ g = 1.65×10^{-3} g

(d) 79 nmol weighs $79 \times 10^{-9} \times 75$ g = 5.93×10^{-6} g

(e) 15 pmol weighs $15 \times 10^{-12} \times 75$ g = 1.13×10^{-9} g

Units of Volume Used in Biochemical Calculations

The liter (L) is used to make buffer solutions and growth media, but to measure most biochemical processes, much smaller volumes are required. The milliliter = 0.001 L or 10^{-3} L; the microliter = 0.000001 or 10^{-6} L.

Problem 1-30

How many microliters are in the following volumes?

(a) 2.83 mL

(b) 0.150 L

(c) 1×10^{-6} L

(d) 0.035 mL

Solution:

(a) 2.83 mL = 2830 μL

(b) 0.150 L = 150,000 μL

(c) 1×10^{-6} = 1 μL

(d) 0.035 mL = 35 μL

Units of Concentration

A *concentration* indicates a ratio of the amount of solute to the amount of solvent or solution. The amount of solute is generally expressed as a unit of weight. The amount of solvent in which the solute is dissolved may be expressed as a weight or a volume. The amount of solution is expressed as a unit of volume. The concentration commonly used in biochemistry are shown in Table 1-5.

Table 1-5. Concentrations Commonly Used in Biochemistry

Concentration Name	Ratio	Units	Abbreviation
Molar	$\dfrac{\text{moles solute}}{\text{liter solution}}$	$\dfrac{\frac{\text{grams solute}}{\text{MW(g)}}}{\text{liter}}$	M
Millimolar	$\dfrac{\text{millimoles}}{\text{liter solution}}$	$\dfrac{\frac{\text{mg solute}}{\text{MW (mg)}}}{\text{liter}}$	mM
Micromolar	$\dfrac{\text{micromoles}}{\text{liter solution}}$	$\dfrac{\frac{\mu\text{g solute}}{\text{MW }(\mu\text{g})}}{\text{liter}}$	μM
Nanomolar	$\dfrac{\text{nanomoles}}{\text{liter solution}}$	$\dfrac{\frac{\text{ng solute}}{\text{MW (ng)}}}{\text{liter}}$	nM
Molal	$\dfrac{\text{moles solute}}{\text{grams solvent}}$	$\dfrac{\frac{\text{grams solute}}{\text{MW (g)}}}{\text{grams solvent}}$	m
Normal	$\dfrac{\text{equivalents}^{a}\text{ of solute}}{\text{liter solution}}$	$\dfrac{\frac{\text{grams solute}}{\text{equivalent wt (g)}}}{\text{liter solution}}$	N
% solutions	wt/wt	$\dfrac{\text{grams solute}}{100\text{ g solvent}}$	%(w/w)
	wt/vol	$\dfrac{\text{grams solute}}{100\text{ mL solution}}$	%(w/v)
		$\dfrac{\text{grams solute}}{\text{decaliter solution}}$	g/dL
		(same as g solute/100 mL solution)	
		$\dfrac{\text{milligrams}}{\text{decaliter solution}}$	mg/dL
Parts per million	$\dfrac{\text{parts solute}}{10^{6}\text{ parts solvent}}$	$\dfrac{\text{milligrams solute}}{\text{liter of solution}^{b}}$	ppm

[a] 1.0 equivalent = the number of moles of solute required to give 1.0 mol H^{+}.
[b] Assuming that 1 mL of solution weighs 1.0 g.

Molarity is the form of concentration most commonly used in biochemistry. *Molality* is used when the physical properties of molecules (e.g., the boiling point or melting point) are being studied since changing temperature does not change the concentration of molal solutions as it does in molar solutions. *Normality* is rarely used with the weak organic acids and bases in biological systems. *Percent solutions* (%) are often used when the molecular weight is unknown. The g/dL notation is frequently used in medicine for the concentrations of compounds, in g/100 mL of serum. *Parts per million* (ppm) is used with gases (e.g., in describing the concentration of air pollutants), but also with liquids in describing the concentration of compounds in rivers and streams.

Problem 1-31

The normal range for glucose concentrations in the human bloodstream is from 80 to 120 mg/dL. Express this in millimolar concentrations (mM). The MW of glucose is 180.

Solution:

$$1 \text{ m}M \text{ glucose} = 180 \text{ mg/L}$$
Then 80 mg/dL = 80 mg/100 mL   or   800 mg/L
$$800 \text{ mg/L} = \frac{800 \text{ mg}/180 \text{ mg (MW)}}{1 \text{ L}} = 4.44 \text{ mmol/L} = 4.44 \text{ m}M$$
$$120 \text{ mg/dL} = 1200 \text{ mg/L} = \frac{1200 \text{ mg}/180 \text{ mg (MW)}}{1 \text{ L}}$$
$$= 6.67 \text{ mmol/L} = 6.67 \text{ m}M$$

Problem 1-32

Express each concentration as a ratio of weight to volume.

(a) 12.25 M glucose

(b) 0.1% glucose solution

(c) 3 ppm glucose in aqueous solution

Solution:

(a) $12.25 \, M = \dfrac{12.25 \text{ mol glucose}}{1 \text{ L solution}}$

(b) $0.1\% = \dfrac{0.1 \text{ glucose}}{100 \text{ mL solution}}$

(c) $3 \text{ ppm} = \dfrac{3 \text{ parts glucose}}{10^6 \text{ parts water}} = \dfrac{3 \text{ μg glucose}}{1 \text{ L water}}$

Problem 1-33

How much alanine must be added to 5 mL of H_2O to give a 3 M solution of alanine?

Solution:

Let the amount of alanine required be x. By definition, 3 M alanine solution is a ratio of 3 mol/L. *Regardless of the volume of the solution, this ratio must hold.* If 5 mL is required, the following proportion may be written

$$\frac{3 \text{ mol}}{1 \text{ L}} = \frac{x}{5 \text{ mL}}$$

$$\frac{3 \text{ mol}}{1.0 \text{ L}} = \frac{x}{0.005 \text{ L}}$$

Solving for x gives us

$$3(0.005) = 1.0x$$
$$x = 0.015 \text{ mol alanine}$$

Problem 1-34

How much glucose must be added to 10.0 mL of a 0.1 mM glucose solution to obtain a 0.25 mM solution?

Solution:

Let the amount of glucose in 10 mL that gives 0.25 mM be x. By definition, 0.25 mM glucose is a ratio of 0.25 mmol/1.0 L. Regardless of the volume of the solution, this ratio must hold and the following proportion may be written:

$$\frac{0.25 \text{ mmol}}{1 \text{ L}} = \frac{x}{0.010 \text{ mL}}$$

$$x = 0.0025 \text{ mmol}$$

A solution of 0.1 mM glucose contains

$$\frac{0.1 \text{ mmol}}{1.0 \text{ L}} \quad \text{or} \quad \frac{0.001 \text{ mmol}}{0.010 \text{ L}}$$

Thus $0.0025 - 0.001 = 0.0015$ mmol of glucose is required.

Problem 1-35 Convert each concentration to moles/liter and express in exponential form.

(a) 75 μM (b) 108 mM

(c) 4.3 nM (d) 8.2 pM

Solution:

(a) 75 μM = 75 $\times$ 10^{-6} M

(b) 108 mM = 108 $\times$ 10^{-3} M or   1.08 $\times$ 10^{-5} M

(c) 4.3 nM = 4.3 $\times$ 10^{-9} M

(d) 8.2 pM = 8.2 $\times$ 10^{-12} M

Problem 1-36

How many micromoles of alanine are in a liver cell (hepatocyte) if the concentration of alanine is 1 mM and the approximate volume of the hepatocyte is 1 $\times$ 10^{-8} mL?

Solution:

$$[\text{ala}] = 1 \text{ m}M = \frac{1 \text{ mmol}}{\text{L}} \text{ or } \frac{1 \text{ }\mu\text{mol}}{\text{mL}}$$

If the volume of the cell = 1 $\times$ 10^{-8} mL or 1 $\times$ 10^{-5} μL, the following ratio must hold:

$$1 \text{ m}M = \frac{1 \text{ }\mu\text{mol}}{1 \text{ mL}} = \frac{x \text{ }\mu\text{mol}}{1 \times 10^{-8} \text{ mL}}$$

$x = 1 \times 10^{-8}$ μmol or 1 $\times$ 10^2 pmol

Problem 1-37

Express each concentration as μmol/mL.

(a) 0.2 mM (b) 9 $\times$ 10^{-6} M

(c) 3.3 $\times$ 10^{-5} M (d) 3.6 mM

Solution:

(a) 0.2 mM = 200 μmol/L or 0.2 μmol/mL

(b) 9 $\times$ 10^{-6} M = 9 μmol/L or 0.009 μmol/mL

(c) 3.3 $\times$ 10^{-5} M = 33 μmol/L or 0.033 μmol/mL

(d) 3.6 mM = 3600 μmol/L or 3.6 μmol/mL

Problem 1-38

(a) How many micromoles of aspartic acid are in 1.0 mL of a 20 μM solution of aspartic acid in water?

(b) How many molecules of aspartic acid are in 1.0 mL of 20 μM aspartic acid?

Solution:

(a) First, find the number of micromoles of aspartate in 1.0 mL of 20 μM solution.

20 μM solution = 20 μmol/L or 20 nmol/mL
1.0 mL of solution contains 20 nmol or 0.02 μmol

(b) By definition, 1 μmol contains 1×10^{-6} mol $\times$ 6.02×10^{23} molecules. Therefore, in 0.02 μmol, there are

$$0.02 \times 10^{-6} \times 6.02 \times 10^{23} = 0.120 \times 10^{(-6+23)}$$
$$= 0.120 \times 10^{17} \text{ molecules}$$

Problem 1-39

The following solutions were made to measure the activity of enzymes in a cell extract. An aliquot of 100 μM glucose solution was added to each tube containing buffer and cell extract. (Assume there is no significant amount of glucose in the cell extract.) Show the number of nanomoles glucose per milliliter in each reaction mixture.

Reaction mixture A	Reaction mixture B
1.0 mL of 100 μM glucose	0.5 mL of 100 μM glucose
1.95 mL of buffer	2.45 mL of buffer
0.05 mL of cell extract	0.05 mL of cell extract
3.0 mL total volume	3.0 mL total volume

Solution:

The total volume is 3.0 mL. In 1.0 mL of a 100 μM solution, there is 100 nmol. In reaction mixture A:

$$\frac{100 \text{ nmol glucose added}}{3 \text{ mL total volume}} = 33.3 \text{ nmol/mL or } 33.3 \text{ } \mu M$$

In 0.5 mL of a 100 μM solution, there is 50 nmol. In reaction mixture B:

$$\frac{50 \text{ nmol glucose added}}{3 \text{ mL total volume}} = 16.7 \text{ nmol/mL or } 16.7 \text{ } \mu M$$

Standard Curves to Determine Concentrations

A standard curve is often used to determine the concentration of a compound in an aqueous solution. Usually, a standard curve relates the absorbance of light by a solution to the concentration of the light-absorbing molecule. For example, compounds containing a benzene ring, such as tyrosine and tryptophan, absorb light in the ultraviolet range of wavelengths (260 to 310 nm). Concentrations of these compounds are easily determined using a spectrophotometer set at the proper wavelengths.

Many biochemical compounds do not absorb light at wavelengths easily measured with simple spectrophotometers. However, these nonabsorbing compounds often react with chemicals to produce derivatives that do absorb light. For example, glucose does not absorb light at easily measured wavelengths, but the aldehyde group of glucose can react with *o*-toluidine to give a colored complex that absorbs light at 630 nm. Under controlled conditions, the amount of *o*-toluidine complex formed is directly proportional to the concentration of glucose in solution. Therefore, measuring the absorption of light at 630 nm, after reaction with *o*-toluidine, allows us to determine the glucose concentration in the original solution.

Spectrophotometry

The absorbance of light by a solution is measured with a spectrophotometer. This instrument causes a beam of incident light of specific wavelength to pass through a sample of the solution, and a photometer measures how much light is transmitted through the sample.

Pure water or solutions that do *not* absorb light transmit 100% of the light that enters the sample. However, solutions that contain light-absorbing molecules will transmit less than 100% of the incident light (I_0) entering the solution (see Fig. 1–1, p.33):

$$T = \frac{I}{I_0} \times 100$$

where T is the percent transmittance, I the transmitted light, and I_0 the incident light.

The I_0 (incident light) is often measured as I (transmitted light) using water or buffer as the sample solution. This is referred to as the *blank* since, theoretically, $I_0 = I$ when no absorbing substance is present in solution. In practice, the blank may have a very small absorbance and the spectrophotometer is then adjusted so that the T of the blank is 100%.

The absorbance or optical density (O.D.) of a solution is inversely related to the transmittance. The *absorbance* is defined as the log of the ratio of incident light (I_0) to transmitted light (I).

$$A = \log \frac{I_0}{I}$$

Beer's Law

The relationship between absorbance and the concentration of the light-absorbing compound is described by *Beer's law*:

$$A = abc$$

where A represents the absorbance or optical density, a the extinction coefficient, b the length of the light path through solution (usually 1 cm), and c the concentration of light-absorbing substance.

Note: The units for a, the extinction coefficient, must be the same as for c.

The extinction coefficient (or absorption coefficient) may be given for different types of concentrations. For example, the *molar extinction coefficient* (a_m or E) is the absorbance of a 1.0 M solution of a particular light-absorbing compound in a 1-cm light path. The *specific absorption coefficient, a,* is used when the concentration is expressed in g/L. The term $a_{1\%}$ is used to denote absorption coefficients when concentrations are g/100 mL.

Problem 1-40

The vitamin riboflavin absorbs light at many different wavelengths (λ). If the molar absorption coefficient is 12,200 at λ = 450 nm, calculate the concentration of riboflavin in each solution, assuming that b = 1 cm.

   (a) Absorbance = 0.11

   (b) Absorbance = 0.91

   (c) Absorbance = 0.34

Solution:

(a) $A = abc$
   $0.11 = 12,200 \times 1 \times c$
   $c = \dfrac{0.11}{12,200} = 9 \times 10^{-6}\ M$

(b) $A = abc$
   $0.91 = 12,200 \times 1 \times c$
   $c = \dfrac{0.91}{12,200} = 74\ \mu M$

(c) $A = abc$
   $0.34 = 12,200 \times 1 \times c$
   $c = \dfrac{0.34}{12,200} = 28\ \mu M$

Problem 1-41

You want a solution of riboflavin that will give you an absorbance of 0.25 O.D. unit. How many millimoles of riboflavin do you add to 0.1 L of water? The a_m = 12,200 for riboflavin and b = 1 cm.

Solution:

Since $A = a_m bc$, and a_m = 12,200, b = 1, and A = 0.25, we may solve for c, the molarity of the solution using Beer's Law.

$$0.25 = 12,200 \times 1 \times c$$

$$c = \frac{0.25}{12,200} = 2 \times 10^{-5} M$$

To obtain 100 mL of $2 \times 10^{-5} M$ riboflavin solution, let moles of riboflavin to be added = x. Then the proportion may be written

$$\frac{0.020 \text{ mmol}}{1.0 \text{ L}} = \frac{x}{0.1 \text{ L}}$$

Solving for x yields

$$x = 0.020 \times 0.1 = 0.002 \text{ mmol} \quad \text{or} \quad 2 \times 10^{-6} \text{ mol} \quad \text{or} \quad 2 \text{ } \mu\text{mol}$$

Problem 1-42

A 0.28 mM solution of the amino acid tyrosine gave an absorbance = 0.37 at a wavelength of 275 nm. Calculate the concentration of tyrosine in a solution that gave an absorbance = 0.09 (assume that the light path is 1 cm).

Solution: For solution 1, let the absorbance be A_1 = 0.37 and [tyrosine] be c_1 = 0.28 mM. For solution 2, let $A = A_2$ = 0.09 and $c = c_2$. From Beer's law, $A = a_m bc$; then $a_m = A/bc$ for each solution. Although the a_m is not given, it must be the same for tyrosine solutions. *Since the extinction coefficients are the same,*

$$a_m = \frac{A_1}{c_1} = \frac{A_2}{c_2}$$

and

$$\frac{0.37}{0.28 \text{ m}M} = \frac{0.09}{c_2}$$

$$c_2 = \frac{0.09 \times 0.28}{0.37} = \frac{0.0252}{0.37} = 0.068 \text{ m}M$$

Problem 1–43

The absorbance spectra are shown in Fig. 1–2 for both the oxidized (NAD^+) and reduced (NADH) forms of the cofactor nicotinamide adenine dinucleotide. Explain which wavelength you would choose to determine [NADH] spectrophotometrically if both NAD^+ and NADH are in the same solution.

Solution:

In selecting the best wavelength to determine [NADH] with NAD^+ in the solution, the spectra of the two forms are compared, as shown in Fig. 1–2. Ideally, one chooses a wavelength where NADH absorbs strongly and NAD^+ not at all. At about 340 nm the absorbance for 1 mM NAD^+ is almost 0, whereas the absorbance for the 0.1 mM NADH solution at 339 nm is 0.62. Thus 339 nm is the wavelength best suited for measuring the [NADH] when NAD^+ is also in solution.

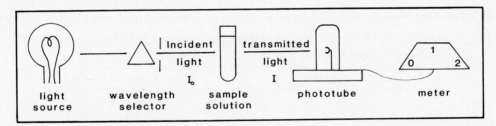

Figure 1–1. Diagram of a simple spectrophotometer. Although the absorbance scale may range from 0 to 2.0, spectrophotometers are most accurate between 0.09 and 1.0.

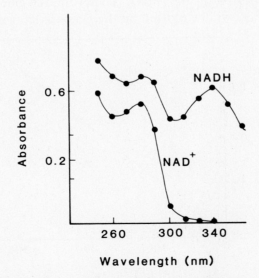

Figure 1–2. Absorbance spectra of the oxidized and reduced forms of NAD.

Constructing a Standard Curve

Determining the concentrations of biological compounds in aqueous solutions is essential to understanding biochemical changes in the cell. A standard curve relates known amounts of a substance in solution to some measurable property such as the absorption of light at a particular wavelength. After plotting absorbances as a function of the compound's concentrations, a relationship may become obvious and a straight line or a curve may be drawn through the points. This line or curve is known as the *standard curve*.

Graphing Data. Each point on a graph has an x value and a y value. The x values are found on the horizontal x-axis or abscissa and the y values are on the vertical y-axis or ordinate. When data points are plotted, a function or relationship between the x and y values may become obvious, and in that case, a smooth line or curve is drawn.

For example, glucose does not absorb light at wavelengths easily measured by a spectrophotometer. Determining [glucose] in solution requires reacting glucose with a compound that will form a light-absorbing substance.

The data in Table 1–6 were obtained using the o-toluidine method for glucose determination on solutions with known concentrations of glucose. This method depends on the aldehyde of glucose reacting with o-toluidine to give a colored complex that absorbs light at 600 nm.

Table 1–6. Absorbances of Toluidine Complexes for Glucose Solutions of Known Concentration

[Glucose] (mg/dL)	Absorbance (600 nm)
50	0.11
100	0.23
150	0.35
200	0.48
300	0.70

In graphing data, the values for the known quantity are usually plotted on the x-axis and the variable quantity is plotted on the y-axis. Therefore, the glucose concentrations will be x values and the absorbances will be plotted on the y-axis as shown in Fig. 1–3.

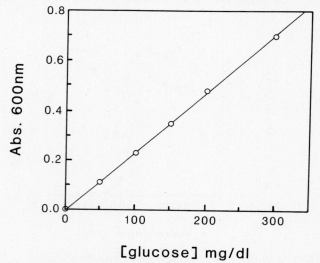

Figure 1–3. Standard curve for the determination of glucose concentration by the o-toluidine method.

Drawing a Line of Best Fit

Since a linear relationship is obvious in this case, a line of best fit is drawn so that the sum of the distance of the points on one side of the line is roughly equal to the sum of the distance of the points on the other side. A more accurate line may be found by using a statistical test, linear regression analysis.

Problem 1-44

Another method for determining glucose concentrations utilizes the coupled reactions of the enzyme glucose oxidase and a reduced chromogen.

$$\text{glucose} + O_2 + H_2O \longrightarrow \text{gluconic acid} + H_2O_2$$
$$\text{reduced chromogen} + H_2O_2 \longrightarrow \text{oxidized chromogen}$$
$$\text{(absorbs light at 420 nm)}$$

The amount of oxidized chromogen is directly proportional to the amount of H_2O_2, which is proportional to the amount of glucose in solution. The concentration of the oxidized chromogen may be determined by reading the absorbance of its solution at 420 nm and consulting a standard curve.

Plot the following data obtained from treating known solutions of glucose with glucose oxidase and reduced chromogen and then reading the absorbances at 420 nm.

Glucose Concentration (μg/mL)	Absorbances of Oxidized Chromagen (O.D. Units at 420 nm)
20	0.08
40	0.15
60	0.22
100	0.39
140	0.51

Solution:

The glucose concentrations are plotted on the *x*-axis since they are the known values, and the absorbances are plotted on the *y*-axis since they are the variable values. From the plot in Fig. 1-4, it is obvious that the relationship is linear, so a line of best fit is drawn.

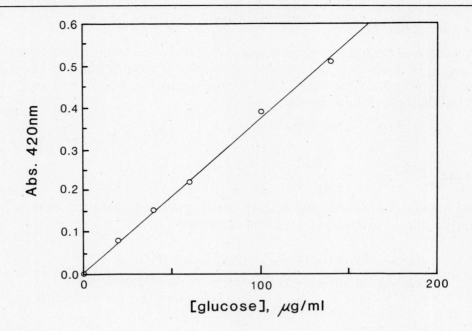

Figure 1-4. Standard curve for glucose concentration by the glucose oxidase method.

Problem 1-45

Normal human urine contains glucose in the range 10 to 130 μg/mL. Normal human blood has a range of glucose concentrations from 0.80 to 1.20 mg/mL. What would these ranges be in absorbance values using the glucose oxidase method?

Solution:

Normal urine: 10 μg/mL = 0.04 O.D. Unit
130 μg/mL = 0.48 O.D. unit

Normal blood: 0.80 mg/mL is 800 μg/mL, which is beyond the limits of accuracy of these graphs. Therefore, dilutions must be made. If 10-fold dilutions are made, then:

1/10 *blood:*   80 μg/mL = 0.30 O.D. unit
120 μg/mL = 0.45 O.D. unit

Problem 1-46

From the standard curve in Fig. 1-3, determine the concentration of glucose in solutions that give each of the following absorbances (O.D. units).

(a) 0.53   (b) 0.05

(c) 0.25   (d) 1.3 (10-fold dilution gave 0.22 O.D. unit)

(e) 1.8 (100-fold dilution gave
    0.15 O.D. unit)

Solution:

(a) 0.53 O.D. unit = 245 mg/dL or 24.5 μg/mL

(b) 0.05 O.D. unit = 20 mg/dL or 2.0 μg/mL

(c) 0.25 O.D. unit = 115 mg/dL or 11.5 μg/mL

(d) The diluted solution is used because the absorbance of the original is too concentrated to be read accurately.

   0.22 O.D. unit = 95 mg/dL or 9.5 μg/mL for the 100-fold dilution
       original = 9.5 $\times$ 10 = 95 μg/mL

(e) Similarly, the diluted solution must be used here as well.

   0.15 O.D. unit = 68 mg/dL or 6.8 μg/mL for the 100-fold dilution
       original = 6.8 $\times$ 100 = 680 μg/mL

Problem 1–47

Calculate each absorbance for the [NADH] given below using Beer's law ($A = abc$). Plot a standard curve for NADH using each [NADH] absorbance.

(a) 0.5 mM

(b) 0.25 mM

(c) 0.20 mM

(d) 0.10 mM

(e) 0.025 mM

Solution:

Using Beer's law, absorbances for each [NADH] may be calculated from $A = a_m bc$, where A represents the absorbance at 340 nm, a_m the *molar* absorptivity coefficient which is 6220 for NADH at λ = 340 nm, b the length of light path through solution = 1 cm, and c the concentration of NADH in mol/L.

(a) A = 6220 $\times$ 1 $\times$ 0.0005 M
    A = 3.1

(b) A = 6220 $\times$ 1 $\times$ 0.00025 M
    A = 1.55

(c) A = 6220 $\times$ 1 $\times$ 0.0002 M
    A = 1.12

(d) A = 0.622

(e) A = 0.155

Plotting these data, we obtain Fig. 1–5.

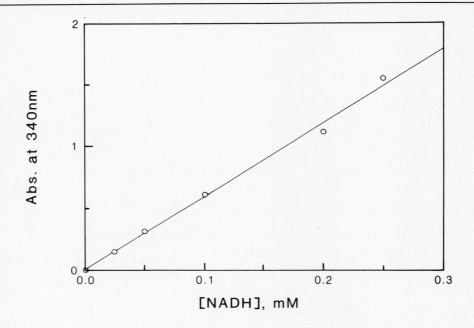

Figure 1-5. Standard curve for [NADH] predicted from the molar extinction coefficient.

Problem 1-48

Using the standard curve plotted in Fig. 1-5, determine the concentration of NADH in each solution.

Solution	Absorbance
1	0.33
2	0.09
3	1.11
4	0.87

Solution:

From the graph in Fig. 1-5, each absorbance is located on the line of best fit and the concentration of NADH at that absorbance is recorded.

Solution	Absorbance	[NADH], mM (μmol/mL)
1	0.33	0.057
2	0.09	0.013
3	1.11	Off-scale; must dilute to read accurately
4	0.87	0.15

In solution 3, the absorbance is beyond the limits of this patricular standard curve and exceeds the limits of accuracy for many spectrophotometers. Therefore, solution 3 must be diluted with water (or buffer) to give an absorbance that can be read accurately on a spectrophotometer.

Standard Curve for Protein Determination by the Lowry Method

Perhaps the standard curve used most often in biochemistry is that for determining the protein concentration in a solution by the Lowry method. Lowry's procedure involves two different color-producing reactions, one in which Cu^+ ions in an alkaline solution react with peptide bonds to form a complex (deep blue) which absorbs light over the range 500 to 660 nm. The second reaction occurs when the inorganic salts in Lowry's reagent react with tryptophan and tyrosine to produce a complex (blue green) that absorbs light over the range 400 to 600 nm. Therefore, the Lowry procedure depends on (1) the presence of peptide bonds, which are always present in any protein, and (2) the tryptophan and tyrosine residues, which are usually, but not always, present in proteins.

The standard curve in Fig. 1-6 was plotted using the absorbance of each solution prepared with a known protein concentration. The accuracy of this method of protein determination is relatively low when protein concentrations are above 2.0 mg/mL.

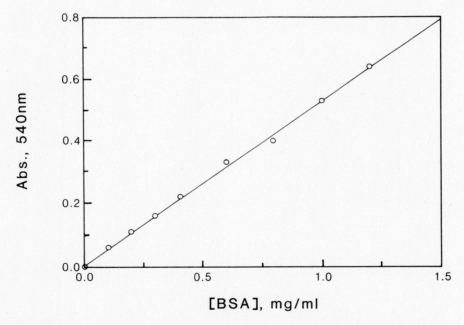

Figure 1-6. Standard curve for Lowry's protein determination. The protein used to construct this curve was albumin from bovine serum. The absorbances were read at 540 nm.

Problem 1-49

The following absorbances (at 540 nm) were read for solutions 1, 2, and 3 after doing the Lowry procedure. Determine the protein concentration for each solution from the Lowry standard curve in Fig. 1-6.

Solution 1: absorbance = 0.35
Solution 2: absorbance = 0.05
Solution 3: absorbance = 0.22 (solution was diluted 10-fold)

Solution:

Solution 1: A_{540} = 0.35, protein concentration = 0.67 mg/mL.
Solution 2: A_{540} = 0.05, protein concentration = 0.08 mg/mL.
Solution 3: The 10-fold dilution has protein concentration = 0.40 mg/mL. The concentration of the original solution 3 was 10 × 0.40 = 4.0 mg/mL.

Exercises

The answers are provided at the back of the book.

1-1. Write each number in exponential form.
(a) 4410
(b) 73,840
(c) 0.0004

1-2. Find the $\log_{10}$ values for each number.
(a) 17
(b) 0.000037
(c) 2,398,451

1-3. Determine the numbers represented by each log value.
(a) $10^{-3.2}$
(b) $10^{-0.5}$
(c) $10^{2.3}$
(d) $\log 10 \doteq 1.6$

1-4. Solve each equation for x.
(a) $16.98x = 4.9 - x$

(b) $-0.01 = -\dfrac{1}{38.5\,(1 + 10/x)}$

(c) $5.1 = 3.9 + \log \dfrac{4.9 - x}{x}$

(d) $x = 6.0 + 1.363 \log \dfrac{(10^{-4})(10^{-7})}{(10^{-2.69})(10^{-4})}$

(e) $0 = -4.5 + 1.363 \log \dfrac{(10^{-3})(10^{-2.72})}{(10^{-x})(10^{-3.69})}$

1-5. The enzyme lactate dehydrogenase catalyzed the formation of lactate at the rate of 0.4 μmol/min. At this rate, how much lactate would be formed in:
(a) 35 sec?
(b) 5.3 min?

1-6. Express each concentration as μmol/mL.
(a) 3.0 mM
(b) 10 μM
(c) 0.1 nM

1-7. How many molecules are in 0.4 μmol of leucine?

1-8. The molecular weight of alanine is 89. Determine how much alanine is required to make each of the following aqueous solutions.
(a) 1.0 L of 1.0 mM alanine
(b) 500 mL of 25 μM alanine
(c) 150 μL of 100 μM alanine
(d) 1.0 L of 50 ppm alanine
(e) 200 mL of 0.02% (w/v) alanine

1-9.   The molar extinction coefficient a_m for tyrosine is 1340 at 275 nm. What would be the absorbance of a tyrosine solution, 150 μM, at 275 nm assuming that the light path, b, was 1.0 cm?

1-10.   Determine the concentration of protein in each solution from the Lowry standard curve in Fig. 1-6.
(a) Absorbance = 0.09
(b) Absorbance = 1.54; 1/100 dilution gave A = 0.12

Chapter 2
ACID-BASE CHEMISTRY

Many biochemical reactions involve acids and bases. There are several different definitions of acids and bases, but the one most frequently used is the Bronsted–Lowry definition.

Bronsted Acids and Bases

An acid or base may be defined in several ways, but the Bronsted acid-base definition is used most often in biochemistry. Bronsted defined acids as proton donors and bases as proton acceptors. Then in the reaction

$$HCl \longrightarrow H^+ + Cl^-$$

HCl is the proton donor and Cl^- may be a proton acceptor. HCl is a strong acid and dissociates completely in water, as indicated by the single reaction arrow.

Weak Acids and Bases

Most acids and bases found in biological systems are weak acids or bases which ionize only partially in an aqueous medium. Partial dissociation is shown by *opposing arrows*, which indicate a dynamic equilibrium between the forward and reverse reactions.

Dissociation of a weak acid

$$\underset{\text{weak acid}}{HA} \rightleftharpoons H^+ + \underset{\text{weak base}}{A^-}$$

Conjugate Acid-Base Pairs

A conjugate acid-base pair is defined as two substances which differ from each other *only* by a proton. In the following reaction, a weak acid, acetic acid, is the proton donor and the acetate ion is the conjugate base.

$$\underset{\text{weak acid}}{CH_3-COOH} \rightleftharpoons \underset{\text{proton}}{H^+} + \underset{\text{conjugate base}}{CH_3-COO^-}$$

A weak base, ammonia (NH_3), accepts a proton to form its conjugate acid, NH_4^+, the ammonium ion.

$$\underset{\text{base}}{NH_3} + \underset{\text{proton}}{H^+} \rightleftharpoons \underset{\text{conjugate acid}}{NH_4^+}$$

Problem 2-1

Show each reactant and product as a conjugate acid-base pair.

(a) $H_2CO_3 \rightleftharpoons H^+ + HCO_3^-$

(b) $CH_3-CHOH-COO^- + H^+ \rightleftharpoons CH_3-CHOH-COOH$

(c) $H_2PO_4^- \rightleftharpoons HPO_4^{2-} + H^+$

Solution:

(a) $H_2CO_3 \rightleftharpoons H^+ + \underset{\text{conjugate base}}{HCO_3^-}$
$\quad \underset{\text{acid}}{}$

(b) $CH_3-CHOH-\underset{\text{base}}{COO^-} + H^+ \rightleftharpoons CH_3-CHOH-\underset{\text{conjugate acid}}{COOH}$

(c) $\underset{\text{acid}}{H_2PO_4^-} \rightleftharpoons H^+ + \underset{\text{conjugate base}}{HPO_4^{2-}}$

Actually, if acetic acid is dissolved in water, there are two different conjugate acid-base pairs in the reaction:

$$CH_3COOH + H_2O \rightleftharpoons CH_3COO^- + H_3O^+$$

(1) $\quad CH_3-COOH \rightleftharpoons H^+ + CH_3-COO^-$

(2) $\quad H_2O + H^+ \rightleftharpoons H_3O^+$

Problem 2-2

Show the following reactants and products as conjugate acid-base pairs.

(a) $H_2CO_3 + H_2O \rightleftharpoons HCO_3^- + H_3O^+$

(b) $H_2PO_4^- + H_2O \rightleftharpoons H_3O^+ + HPO_4^{2-}$

Solution:

(a) $\underset{\text{weak acid}}{H_2CO_3} + \underset{\text{weak base}}{H_2O} \rightleftharpoons \underset{\text{conjugate acid}}{H_3O^+} + \underset{\text{conjugate base}}{HCO_3^-}$

(b) $\underset{\text{weak acid}}{H_2PO_4^-} + \underset{\text{weak base}}{H_2O} \rightleftharpoons \underset{\text{conjugate acid}}{H_3O^+} + \underset{\text{conjugate base}}{HPO_4^{2-}}$

Note: H^+ ions exist in aqueous solutions as H_3O^+ (hydronium ions) but are written H^+ for simplicity.

Dissociation Constants Are Equilibrium Constants

Reversible reactions proceed in both forward and reverse directions simultaneously but generally at different rates. In the reaction

$$A \underset{k_2}{\overset{k_1}{\rightleftharpoons}} B + C$$

the rate of the forward reaction $(A \rightarrow B + C)$ is $k_1[A]$, and the rate of the reverse reaction $(B + C \rightarrow A)$ is $k_2[B][C]$, where k_1 and k_2 are rate constants.

If the system is allowed to come to equilibrium, then by definition the forward rate = the reverse rate. Then $k_1[A] = k_2[B][C]$ at equilibrium; rewriting this expression gives us

$$\frac{k_1}{k_2} = \frac{[B][C]}{[A]}$$

The ratio of k_1/k_2 under equilibrium conditions is called the equilibrium constant, K_{eq}.

Thus

$$K_{eq} = \frac{k_1}{k_2} = \frac{[B][C]}{[A]}$$

Note: K_{eq} is expressed as a constant without units.

When the reaction being described is the dissociation of a weak acid,

$$HA \rightleftharpoons H^+ + A^-$$

then K_{eq} is designated K_a, the *dissociation constant.*

Since $K_a = k_1/k_2 = [H^+][A^-]/[HA]$, then K_a may be determined by measuring the *molar* concentrations HA, A^-, and H^+ ion at equilibrium. Like K_{eq}, K_a is expressed as a number without units. Weak acids dissociate incompletely, and the reverse reaction predominates so that much more of the compound resides in the acid form, HA, than in its conjugate base, A^-. Therefore, the dissociation constant for a weak acid is very small, generally less than 1×10^{-3}. Some commonly used dissociation constants are listed in Table 2-1 (p. 46).

Problem 2-3

Calculate the K_a for the dissociation of the weak acid, $HA \rightleftharpoons H^+ + A^-$, in an aqueous solution, where $[H^+] = 2 \ \mu M$, $[A^-] = 2 \ \mu M$, and $[HA] = 0.02 \ mM$.

Solution:

$$K_a = \frac{[H^+][A^-]}{[HA]}$$

Converting each to molar concentration, we have

$$K_a = \frac{(2 \times 10^{-6} \ M)(2 \times 10^{-6} \ M)}{(2 \times 10^{-5} \ M)}$$

$$= \frac{4 \times 10^{-12} \ M}{2 \times 10^{-5} \ M}$$

$$= 2 \times 10^{-7}$$

Note: $2 \ \mu M \times 2 \ \mu M$ is *not* 4 μM. Each must be converted to $2 \times 10^{-6} \ M$ before multiplying.

Table 2-1. Dissociation Constants for Some Weak Acids in Biological Systems

Weak acid (HA)	Conjugate Base (A⁻)	K_a
CH_3-COOH (acetic acid)	CH_3-COO^- (acetate)	1.82×10^{-5}
$CH_3-CHOH-COOH$ (lactic acid)	$CH_3-CHOH-COO^-$ (lactate)	1.38×10^{-4}
$HCOOH$ (formic acid)	$HCOO^-$ (formate)	1.76×10^{-5}

Problem 2-4

Calculate the [acetic acid] in an aqueous solution where [acetate] = 0.4 mM and [H⁺] = 3.9×10^{-4} M.

Solution:

$$CH_3-\underset{\text{HA}}{COOH} \rightleftharpoons H^+ + \underset{A^-}{CH_3-COO^-}$$

From Table 2-1, the K_a for acetic acid is 1.82×10^{-5}.

$$K_a = 1.82 \times 10^{-5} = \frac{[H^+][A^-]}{[HA]} = \frac{[CH_3COO^-][H^+]}{[CH_3COOH]}$$

Let [HA] = [CH₃COOH]. Substitituing yields

$$[A^-] = 4 \times 10^{-4} \, M$$

$$[H^+] = 3.9 \times 10^{-4} \, M$$

Then

$$1.82 \times 10^{-5} \, M = \frac{(4 \times 10^{-4} \, M)(3.98 \times 10^{-4} \, M)}{[HA]}$$

$$[HA] = \frac{15.9 \times 10^{-8} \, M}{1.8 \times 10^{-5} \, M} = 8.8 \times 10^{-3} \, M \quad \text{or} \quad 8.8 \, \text{m}M$$

Thus the [acetic acid] is 8.8 mM.

Problem 2-5

Which of the following is the stronger acid?

Lactic acid: $K_a = 1.38 \times 10^{-4}$

Acetic acid: $K_a = 1.82 \times 10^{-5}$

Solution:

The rule to follow is: The larger the K_a, the stronger the acid. Since $K_a = [H^+][A^-]/[HA]$, the larger K_a indicates that there was greater dissociation of the acid. In this case, lactic acid is the stronger acid.

Problem 2-6

The dissociation constant, K_a, for formic acid (HCOOH) is 1.76×10^{-4}. Calculate [formate] in a solution containing 100 μM H^+ ions and 28 μM formic acid.

Solution:

Formic acid (HA) partially dissociates to give H^+ and formate, the conjugate base.

$$\underset{(HA)}{HCOOH} \rightleftharpoons H^+ + \underset{(A^-)}{HCOO^-}$$

$$K_a = 1.76 \times 10^{-4} = \frac{[H^+][HCOO^-]}{[HCOOH]}$$

Let $[HCOO^-] = A^-$; then substituting in the equation above yields

$$1.76 \times 10^{-4} = \frac{(1 \times 10^{-4}\,M)[A^-]}{2.8 \times 10^{-5}}$$

Rearranging the equation gives

$$\frac{(1.76 \times 10^{-4}\,M)(2.8 \times 10^{-5})}{1 \times 10^{-4}} = [A^-]$$

$$[A^-] = \frac{4.93 \times 10^{-9}}{1 \times 10^{-4}}$$

$$= 4.9 \times 10^{-5}\,M = [HCOO^-]$$

so [formate] is $4.9 \times 10^{-5}\,M$.

pH

The H^+ ion concentration is generally a very small number and it is more convenient to express it as its logarithm. The term *pH* is defined as the negative $\log_{10}$ of the $[H^+]$ in moles/liter, or

$$pH = -\log[H^+]$$

In general, the pH scale is used as a measure of acidity. The pH scale ranges from 0 to 14. Strong acids give solutions of low pH since they dissociate proportionately more H^+ ions (e.g., 1 N HCl dissociates completely to give $[H^+] = 0.1\,M$; then pH $= -\log 0.1 = -\log 10^{-1} = 1$). A neutral pH, around 7, is found in most biological systems and indicates that the $[H^+]$

is about $1 \times 10^{-7} M$. A solution of very strong base may have a pH = 13 since the OH^- ions dissociated from the base combine with the H^+ ions in the water, decreasing the $[H^+]$ to $1 \times 10^{-13} M$.

Problem 2-7

Note: It may be helpful to consult Chapter 1 on logarithms.

Determine the pH of each solution.

(a) A cola drink where normally $[H^+] = 0.0008\ M$

(b) A household ammonia cleaning solution where $[H^+] = 1 \times 10^{-12}\ M$

(c) Laboratory distilled water, $[H^+] = 1.2 \times 10^{-6}\ M$

(d) Pasteurized milk, where $[H^+] = 2.7 \times 10^{-7}\ M$

Solution:

(a) $pH = -\log(0.0008) = -\log 10^{-3.09} = -(-3.09) = 3.09$

(b) $pH = -\log(1 \times 10^{-12}) = -\log 10^{-12} = -(-12) = 12$

(c) $pH = -\log(1.2 \times 10^{-6}) = -\log 10^{-5.9} = -(-5.9) = 5.9$

(d) $pH = -\log(2.7 \times 10^{-7}) = -\log 10^{-6.7} = -(-6.7) = 6.7$

Henderson-Hasselbalch Equation

The Henderson–Hasselbalch equation relates the K_a to the pH of a solution of a weak acid.

Since

$$K_a = \frac{[H^+][A^-]}{[HA]}$$

the terms may be rearranged so that

$$[H^+] = \frac{K_a[HA]}{[A^-]}$$

Taking the logarithms of both sides gives us

$$\log[H^+] = \log K_a + \log\frac{[HA]}{[A^-]}$$

Multiplying both sides by -1 yields

$$-\log [H^+] = -\log K_a - \log \frac{[HA]}{[A^-]}$$

$-\log K_a$ is defined as pK_a.

Since $-\log [H^+] = pH$ and $-\log K_a = pK_a$, the equation above may be rewritten

$$pH = pK_a + \log \frac{[A^-]}{[HA]}$$

where $[A^-]$ = [conjugate base] and $[HA]$ = [conjugate acid]. This equation is known as the *Henderson–Hasselbalch equation.*

Note: $pH = pK$ when $[A^-] = [HA]$ since $\log([A^-]/[HA]) = \log(1) = 0$, and $pH = pK + 0$; $pH = pK$.

Problem 2–8

Calculate the pK_a values for each weak acid from its K_a value given in Table 2-1 (p. 46).

(a) Acetic acid

(b) Lactic acid

(c) Formic acid

Solution:

(a) $pK_a = -\log K_a = -\log 1.82 \times 10^{-5} = -\log 10^{-4.74}$
$= 4.74$

(b) $pK_a = -\log K_a = -\log 1.38 \times 10^{-4} = -\log 10^{-3.86}$
$= 3.86$

(c) $pK_a = -\log K_a = -\log 1.76 \times 10^{-5} = -\log 10^{-4.75}$
$= 4.75$

Problem 2–9

Calculate the pH of a solution (0.5 L) containing 1.0 mmol of formic acid and 0.4 mmol of formate ion. The pK_a of formic acid is 4.75.

Solution:

The pH of the solution may be determined using the Henderson–Hasselbalch equation,

$$pH = pK_a + \log \frac{[A^-]}{[HA]}$$

where $[A^-]$ = [formate] and $[HA]$ = [formic acid].

$$[A^-] = 0.4 \text{ mmol}/0.5 \text{ L} = 0.8 \text{ m}M \quad \text{or} \quad 8 \times 10^{-4} M$$

$$[HA] = 1.0 \text{ mmol}/0.5 \text{ L} = 2.0 \text{ m}M \quad \text{or} \quad 2 \times 10^{-3} M$$

$$pH = 4.75 + \log \frac{8 \times 10^{-4}}{2 \times 10^{-3}}$$

$$= 4.75 + \log (4 \times 10^{-1}) = 4.75 + \log (10^{-0.39})$$

$$= 4.75 + (-0.39)$$

$$= 4.36$$

Problem 2-10

Determine the ratio of the concentration of acetate to acetic acid when the pH of the solution is 5 and the K_a of acetic acid is 1.81×10^{-5}.

Solution:

The ratio of [acetate] to [acetic acid] may be determined using the Henderson–Hasselbalch equation. First determine the pK_a of acetic acid by taking the $-\log$ of the K_a:

$$-\log (1.81 \times 10^{-5}) = 4.74$$

Now determine the ratio of acetate to acetic acid.

$$pH \quad = pK + \log \frac{[\text{acetate}]}{[\text{acetic acid}]}$$

$$5.0 \quad = 4.74 + \log \frac{[\text{acetate}]}{[\text{acetic acid}]}$$

$$0.26 \quad = \log \frac{[\text{acetate}]}{[\text{acetic acid}]} \quad \text{antilog of } 0.26 = 1.82$$

$$1.82 \quad = \frac{[\text{acetate}]}{[\text{acetic acid}]}$$

Therefore,

$$\frac{[\text{acetate}]}{[\text{acetic acid}]} = \frac{1.82}{1}$$

Problem 2-11

The bicarbonate ion dissociates to form the carbonate ion and a proton:

$$HCO_3^- \rightleftharpoons CO_3^{2-} + H^+$$

bicarbonate   carbonate
(weak acid)   (conjugate
               base)

Given that the pK_a for this conjugate acid-base pair is 10.3, what is the pH of a solution in which the ratio of carbonate to bicarbonate is 2 to 1?

Solution:

Use the Henderson–Hasselbalch equation to solve for pH:

$$pH = pK_a + \log \frac{[CO_3^{2-}]}{[HCO_3^-]}$$

Substitute the ratio of $[CO_3^{2-}]/[HCO_3^-] = 2/1$:

$$pH = 10.3 + \log \frac{2}{1} \qquad \log \text{ of } 2 = 0.3$$

$$= 10.3 + 0.3$$

$$= 10.6$$

Problem 2-12

A weak acid is in solution at pH 6.0 and the ratio of the concentration of the weak acid to its conjugate base is 3:1. What is the value of K_a?

Solution:

The compound may exist in solution as

$$HA \rightleftharpoons A^- + H^+$$

weak acid   conjugate base

Use the Henderson–Hasselbalch equation to solve for pK_a:

$$pH = pK_a + \log \frac{[A^-]}{[HA]}$$

Substitute:

$$6.0 \;\; = pK_a + \log \frac{1}{3}$$

$$= pK_a + (-0.48)$$

$$6.48 \;\; = pK_a$$

Since $pK_a = -\log K_a$,

$$-\log K_a = 6.48$$

$$\log K_a = -6.48$$

$$K_a = \text{antilog of } -6.48$$

$$= 3.31 \times 10^{-7}$$

Problem 2-13

Calculate the pK_a for pyruvic acid given the following data. In a solution, pH = 4.70, [pyruvic acid] = 0.005 M, and [pyruvate ion] (unprotonated form) = 0.035 M.

Pyruvic acid has one dissociable proton per molecule.

Solution:

Use the Henderson–Hasselbalch equation:

$$pH = pK + \log \frac{[A^-]}{[HA]}$$

We may substitute values given above.

$$4.70 = pK_a + \log \frac{[0.035\ M]}{[0.005\ M]}$$

$$= pK_a + \log 7.0$$

The log of 7.0 = 0.845 (i.e., 7.0 = $10^{0.845}$)

$$4.70 = pK_a + 0.845$$

$$pK_a = 4.70 - 0.845$$

$$= 3.85$$

Problem 2-14

If 0.55 g of sodium lactate (MW = 112) is dissolved in 1.0 L of water at pH 5.1, what will be the concentration of the protonated form, lactic acid? Assume that the pK for the carboxyl group is 3.87.

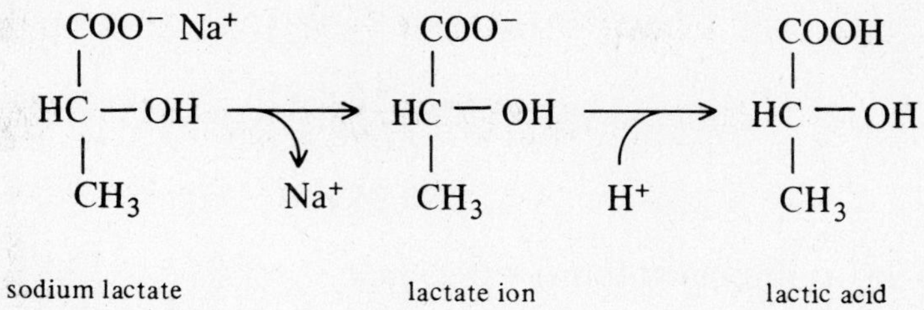

$$
\begin{array}{ccc}
\text{COO}^- \text{Na}^+ & \text{COO}^- & \text{COOH} \\
| & | & | \\
\text{HC}-\text{OH} \longrightarrow & \text{HC}-\text{OH} \longrightarrow & \text{HC}-\text{OH} \\
| \qquad\quad \searrow & | \qquad\quad \nearrow & | \\
\text{CH}_3 \qquad \text{Na}^+ & \text{CH}_3 \qquad \text{H}^+ & \text{CH}_3
\end{array}
$$

sodium lactate   lactate ion   lactic acid

Note: Sodium lactate molecules dissociate completely in water to give lactate ions (unprotonated form), some of which pick up protons to become lactic acid (protonated form).

Solution:

First, determine the total concentration of lactate (unprotonated) and lactic acid (protonated) forms in solution.

$$\frac{0.55\ \text{g}}{122} = 0.0045\ \text{mol} \quad \text{or} \quad 4.5\ \text{mmol}$$

$$\frac{4.5\ \text{mmol}}{1.0\ \text{L}} = 4.5\ \text{m}M$$

Let [protonated form, lactic acid] = x mM, and let [unprotonated form, lactate ion] = 4.5 m$M - x$, since these are the only two forms in solution. Substituting in the Henderson-Hasselbalch equation,

$$\text{pH} = \text{p}K + \log \frac{[\text{A}^-]}{[\text{HA}]}$$

we have

$$5.1 = 3.87 + \log \frac{4.5 - [x]}{[x]}$$

Solving for x gives

$$1.23 = \log \frac{4.5 - [\text{HA}]}{[\text{HA}]}$$

The antilog of 1.23 = 16.98, so

$$16.98 = \frac{4.5 - [\text{HA}]}{[\text{HA}]}$$

$$16.98[\text{HA}] = 4.5 - [\text{HA}]$$

$$17.98[HA] = 4.5$$

$$[HA] = \frac{4.5}{17.98}$$

$$= 0.25 \text{ m}M$$

Thus the [lactic acid] (protonated form) = 0.25 mM.

Problem 2-15

If 7.8 mg of the salt $K^+H_2PO_4^-$ (MW = 132) is dissolved in 500 mL of water at pH 6.5, what will be the final concentration of HPO_4^{2-}, the unprotonated form? Assume that the pK for $H_2PO_4^-$ dissociation is 6.86.

$$H_2O + KH_2PO_4 \xrightarrow{\quad K^+ \quad} \underset{HA}{H_2PO_4^-} \rightleftharpoons \underset{A^-}{HPO_4^{2-}} + H^+$$

Solution:

First, determine the total concentration of protonated ($H_2PO_4^-$) and unprotonated (HPO_4^{2-}) forms in solution. Since 1.0 mmol of KH_2PO_4 weighs 132 mg,

$$\frac{7.8 \text{ mg}}{132 \text{ mg}} = 0.059 \text{ mmol}$$

Let the concentration of HA + A^- be x,

$$\frac{0.059 \text{ mmol}}{500 \text{ mL}} = \frac{x}{1.0 \text{ L}}$$

$$\frac{0.059 \text{ mmol}}{0.5 \text{ L}} = \frac{x}{1.0 \text{ L}} = \frac{0.118 \text{ mmol}}{1.0 \text{ L}} = 0.118 \text{ m}M = [HA] + [A^-]$$

Let $[HPO_4^{2-}]$, the unprotonated form, be $[A^-]$. Since $[A^-] + [HA] = 0.118$ mM, then $[H_2PO_4^-] = [HA] = 0.118 - [A^-]$. Substituting in the Henderson-Hasselbalch equation gives us

$$pH = pK + \log \frac{[A^-]}{[HA]}$$

$$6.51 = 6.86 + \log \frac{[A^-]}{0.118 - [A^-]}$$

Solving for x, we have

$$6.51 - 6.86 = \log \frac{[A^-]}{0.118 - [A^-]}$$

$$-0.35 = \log \frac{[A^-]}{0.118 - [A^-]}$$

The antilog of -0.35 is 0.44. Then

$$\text{antilog } -0.35 = 0.44 = \frac{[A^-]}{0.118 - [A^-]}$$

$$0.44(0.118 - [A^-]) = [A^-]$$

$$0.052 - 0.44[A^-] = [A^-]$$

$$1.44[A^-] = 0.052$$

$$[A^-] = 0.036 \text{ m}M$$

Thus $[HPO_4{}^{2-}] = 0.036$ mM.

Titration Curves

A *titration curve* is a graphical representation of the pH changes that occur during the addition of H^+ or OH^- ions to a solution. To construct a titration curve, known amounts of acid or base are added to the solution being titrated. The added amounts of acid or base are plotted on the x-axis. After each addition, the pH is measured with a pH meter and plotted on the y-axis. The x-axis may also be plotted in *equivalents* of acid or base. In titrations, *an equivalent is the amount of base required to dissociate completely all the protons from an acid form in solution,* or the amount of acid required to protonate a base completely.

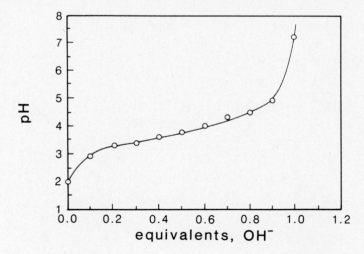

Figure 2–1. Titration curve for lactic acid titrated with NaOH.

The titration curve for acetic acid in Fig. 2-1 demonstrates the *plateau of buffering* seen when the pH of the solution is within 1 pH unit on either side of the pK_a value.

Consider the dissociation of protons from a weak acid, HA, during the addition of strong base, NaOH.

$$NaOH \longrightarrow Na^+ + OH^-$$

$$HA \rightleftharpoons H^+ + A^-$$

After the addition of 1.0 equivalent of base (OH^-), all the acid should be in its conjugate base form, A^-, and the pH more than 2 pH units above the pK. At the point during the titration where 0.5 equivalent of base has been added, one half of the acid should be dissociated and the other half still HA. Then at 0.5 equivalent,

$$[HA] = [A^-] \quad \text{and} \quad \frac{[A^-]}{[HA]} = 1.0$$

From the Henderson–Hasselbalch equation,

$$pH = pK_a + \log 1.0$$

Since the log of 1.0 is 0, $pH = pK_a$ at 0.5 equivalent of base added.

Problem 2-16

Using the Henderson–Hasselbalch equation, show each of the following for the reversible dissociation of the weak acid,

$$HA \rightleftharpoons H^+ + A^-$$

(a) $pH = pK_a$ when $[HA] = [A^-]$

(b) pH when $[HA] = 10[A^-]$

(c) pH when $[HA] = 100[A^-]$

(d) pH when $[A^-] = 100[HA]$

Solution:

$$pH = pK_a + \log \frac{[A^-]}{[HA]} \quad \text{Henderson–Hasselbalch equation}$$

(a) When $[HA] = [A^-]$, $[A^-]/[HA] = 1.0$.

$$pH = pK_a + \log (1.0) \quad \log (1.0) = 0$$

$$= pK_a + 0$$

Thus $pH = pK_a$ when $[A^-] = [HA]$.

(b) When $[HA] = 10[A^-]$, $[A^-]/[HA] = 1/10 = 0.1$.

$pH = pK_a + \log (0.1)$ $\log (0.1) = -1$

$\quad = pK_a - 1$

Thus the pH is 1 pH unit below the pK_a when $[HA] = 10[A^-]$.

(c) When $[HA] = 100[A^-]$, $[A^-]/[HA] = 1/100 = 0.01$.

$pH = pK_a + \log (0.01)$ $\log (0.01) = -2$

$\quad = pK_a - 2$

Thus the pH is 2 pH units below pK_a when $[HA] = 100[A^-]$.

(d) When $[A^-] = 100[HA]$, $[A^-]/[HA] = 100/1$.

$pH = pK_a + \log (100)$ $\log 100 = 2$

$\quad = pK_a + 2$

Thus the pH is 2 pH units above the pK_a when $[A^-] = 100[HA]$.

Problem 2-17

Plot the following titration data obtained during the titration of 100 mL of 0.1 M acetic acid with 0.05 N NaOH.

mL of 0.05 N NaOH Added	pH
0	2.9
10	3.5
20	3.8
30	4.1
40	4.2
50	4.3
60	4.3
70	4.4
80	4.5
90	4.6
100	4.7
110	4.8
120	4.9
130	5.0
140	5.2
150	5.3
160	5.4
170	5.6
180	5.8
190	6.5
200	8.1

Solution:

The pH values are plotted on the y-axis and the volume (mL) of 0.05 M NaOH added is plotted on the x-axis; a smooth curve may be drawn through the points showing the plateau of buffering (Fig. 2–2).

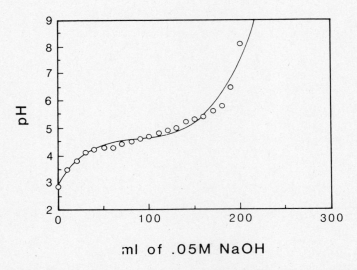

Figure 2–2. Titration data plotted as pH versus volume of NaOH added to 0.1 M acetic acid solution.

Problem 2–18

From the titration curve in Fig. 2–2, determine:

(a) Moles of NaOH equal to 1 equivalent

(b) pK_a for acetic acid (CH_3–COOH)

Solution:

(a) From the titration curve in Fig 2–2., approximately 200 mL of 0.05 N NaOH is required to titrate completely 100 mL of 0.1 M CH_3COOH. Since NaOH is monobasic, 0.05 N NaOH = 0.05 M NaOH. 200 mL of 0.05 M NaOH contains

200 mL $\times$ 0.05 mol/1000 mL = 0.01 mol

(b) pK_a = pH at 0.5 equivalent. Since 1.0 equivalent = 100 mL, 0.5 equivalent = 100 mL. After the addition of 100 mL, pH = 4.7. Therefore, pK_a = 4.7.

Problem 2–19

Using the Henderson–Hasselbalch equation, show that:

(a) The initial solution of 0.1 M acetic acid at pH 2.9 in Problem 2–17 contains almost 100% CH_3–COOH.

(b) The final solution, 100 mL of 0.1 *M* acetic acid plus 200 mL of 0.05 *N* NaOH, contains almost 100% acetate (conjugate base).

Solution:

(a) pH = pK_a + log [A$^-$]/[HA], where [HA] = [acetic acid] and [A$^-$] = [acetate ion]. Substituting the pH and pK_a values (determined in Problem 2–18) gives us

$$2.9 = 4.7 + \log \frac{[A^-]}{[HA]}$$

$$-1.8 = \log \frac{[A^-]}{[HA]}$$

$$0.016 = \frac{[A^-]}{[HA]}$$

$$[A^-] = 0.016 \times [HA]$$

Then [A$^-$] = less than 2% [HA].

(b) After 200 mL of 0.05 *M* NaOH was added, the pH rose to 8.1. Then 8.1 = 4.7 + log ([A$^-$]/[HA]):

$$3.4 = \log \frac{[A^-]}{[HA]}; \quad \text{antilog } 3.4 = 2511 = \frac{[A^-]}{[HA]}$$

Thus [A$^-$] is 2511 times [HA] at pH 8.1.

Buffers

Buffers are compounds that resist changes in the pH of a solution. Most effective buffer systems consist of a weak acid (HA) and the salt of its conjugate base (X$^+$A$^-$). Each dissociates to give the common ion, A$^-$. Consider the acetic acid:Na acetate (CH$_3$–COOH: CH$_3$–COO$^-$Na$^+$) buffer system. CH$_3$–COOH is a weak acid that dissociates incompletely to give a proton and acetate, the conjugate base.

$$CH_3-COOH \rightleftharpoons CH_3-COO^- + H^+$$

CH$_3$–COO$^-$Na$^+$, the salt of the weak acid, dissociates completely to give a Na$^+$ ion and acetate.

$$CH_3-COO^-Na^+ \longrightarrow CH_3-COO^- + Na^+$$

Acetate, CH$_3$–COO$^-$, *is the common ion in this buffer system.*

Note: The salt will dissociate completely, but the weak acid will not.

There is then a reservoir of conjugate base (the common ion) available to buffer protons added to the solution from other chemical reactions. Since weak acids ionize incompletely, there is also a reservoir of protons in the undissociated weak acid to buffer added OH^- or, more commonly, to replace protons that are consumed in biochemical reactions.

Buffers are most effective when the pH is at the pK_a of the weak acid since at this pH [proton donor] = [proton acceptor]. Added H^+ ions will associate with the conjugate base and there is little change in the pH. On the other hand, if OH^- ions are added, the H^+ ions dissociate from the weak acid and combine with the OH^- to give H_2O. Then no significant increase in pH occurs with the addition of these H^+ or OH^- ions until the capacity of the buffer system has been exceeded. In many biochemical reactions, H^+ ions are removed from solution, causing a potential increase in pH if the solution is unbuffered.

Problem 2-20

Write the chemical reactions describing the dissociation of each buffer component and identify the common ion for each system.

(a) $H_2CO_3 + Na^+HCO_3^-$; the bicarbonate buffer system

(b) $HCOOH + HCOO^-Na^+$; the formate buffer system

Solution:

(a) $H_2CO_3 \rightleftharpoons H^+ + HCO_3^-$

$Na^+HCO_3^- \longrightarrow Na^+ + HCO_3^-$; HCO_3^- is the common ion

(b) $HCOOH \rightleftharpoons H^+ + HCOO^-$

$Na^+HCOO^- \longrightarrow Na^+ + HCOO^-$; $HCOO^-$ is the common ion

Problem 2-21

A buffer solution is formed by combining equal volumes of 0.1 M HA and 0.1 M Na^+A^-. The final pH is 6.5 and the pK_a for HA is 6.0.

(a) Determine the ratio of $[A^-]/[HA]$ in this solution.

(b) Determine the $[A^-]$ and $[HA]$ under these conditions.

Solution:

(a) The buffer components dissociate as follows:

$$HA \rightleftharpoons A^- + H^+$$
(0.1 M)   (x)   (x)

$$Na^+A^- \longrightarrow Na^+ + A^-$$
(0.1 M)   (0.1 M) (0.1 M)

Then, assuming that $[HA] = 0.1\ M - x$ and $[A^-] = 0.1\ M + x$,

$$6.5 = 6.0 + \log \frac{0.1 + x}{0.1 - x}$$

$$0.5 = \log \frac{0.1 + x}{0.1 - x} \; ; \quad \text{antilog } 0.5 = 3.16$$

$$3.16 = \frac{0.1 + x}{0.1 - x}$$

Thus the ratio of $[A^-]/[HA]$ is approximately 3.16 at pH 6.5.

(b) To determine the $[A^-]$ and $[HA]$, solve the equation in (a) for x; $3.16(0.1 - x) = 0.1 + x$

$$-3.16x + 0.316 = 0.1 + x \; ; \quad 4.16x = 0.216; \quad x = 0.05 \; M$$

Thus $[A^-] = 0.1 + 0.05 = 0.15 \; M$ and $[HA] = 0.10 - 0.05 = 0.05 \; M$.

Problem 2-22

Tris buffers contain the weak base tris (hydroxymethyl) aminomethane, MW = 121 and $pK_a = 8.3$. In aqueous solution, Tris base ($Tris^0$) molecules pick up protons to give the conjugate acid form, $Tris^+$. In a 0.01 M Tris buffer solution, pH 7.8, how much Tris exists as $Tris^0$ and how much as $Tris^+$?

Solution:

Using the Henderson–Hasselbalch equation, the ratio $Tris^0/Tris^+$ can be found.

$$7.8 = 8.3 + \log \frac{[Tris^0]}{[Tris^+]}$$

$$-0.5 = \log \frac{[Tris^0]}{[Tris^+]} \; ; \quad \text{antilog } -0.5 = 0.316$$

$$0.316 = \frac{[Tris^0]}{[Tris^+]}$$

Since $[Tris^0] + [Tris^+] = 0.01 \; M$, then $[Tris^0] = 0.01 - [Tris^+]$. Substituting in the expression $0.316 = [Tris^0]/[Tris^+]$, we obtain

$$0.316 = 0.01 - [Tris^+]$$

$$0.316[Tris^+] = 0.01 - [Tris^+]$$

$$1.316[Tris^+] = 0.01$$

$$[Tris^+] = 0.0076 \; M$$

$$[Tris^0] = 0.01 - 0.0076 = 0.0024 \; M$$

Problem 2-23

To maintain a solution at pH 6.8 a phosphate buffer is selected in which the weak acid is $H_2PO_4^-$ ($pK_a = 6.86$) and the conjugate base is HPO_4^{2-}. The sodium salts of these two components were used to make a buffer solution, pH 6.76.

(a) Write the chemical reactions showing the dissociation of the two buffer components.

(b) Identify the common ion in this buffer system.

(c) Assuming that 0.2 M solutions of each component are available, what volumes of each solution must be combined to give a final solution with pH 6.76?

Solution:

(a) $NaH_2PO_4 \longrightarrow Na^+ + H_2PO_4^- \rightleftharpoons H^+ + HPO_4^{2-}$.

$Na_2HPO_4 \longrightarrow 2Na^+ + HPO_4^{2-}$

(b) The weak acid is $H_2PO_4^{2-}$ and its conjugate base is HPO_4^-.

(c) Combining equal volumes of 0.2 M solutions, A^- and HA, gives a total buffer concentration $[A^-] + [HA] = 0.4 \ M$. Let $[HPO_4^{2-}] = [A^-] = 0.4 \ M - [HA]$. Using the Henderson–Hasselbalch equation to determine the ratio of $[A^-]/[HA]$.

$$6.76 = 6.86 + \log \frac{[A^-]}{[HA]}$$

$$-0.1 = \log \frac{[A^-]}{[HA]}; \quad \text{antilog of } -0.1 = 0.79$$

$$0.79 = \frac{[A^-]}{[HA]}$$

Assuming that HA is formed mainly by the NaH_2PO_4 and A^- mainly by Na_2HPO_4, these buffer components are added in the ratio 0.79.

Since the molarities are identical, we may use the ratio of volumes of the component solutions. Then

$$\frac{mL \ 0.2 \ M \ Na_2HPO_4}{mL \ 0.2 \ M \ Na_2H_2PO_4} = 0.79$$

$$mL \ A^- = 0.79 \times mL \ HA$$

Then to make up 100 mL of this buffer at pH 6.76,

$$0.79 = \frac{mL \ A^-}{100 \ mL - mL \ HA}$$

$$mL \ A^- = 44; \quad mL \ HA = 100 - 44 = 56$$

Thus 44 mL of 0.2 M Na_2HPO_4 and 56 mL of 0.2 M NaH_2PO_4 should give 100 mL of 0.4 M buffer at pH 6.7.

Problem 2-24

Explain how 1.0 L of the following buffers are made.

(a) Tris buffer, 0.02 M, pH 8.0 (MW Tris = 121; pK = 8.3)

(b) NaH_2PO_4:Na_2HPO_4 buffer, 0.2 M, pH 7.0 (MW NaH_2PO_4 = 122; MW Na_2HPO_4 = 142); pK 6.86

Solution:

(a) To make 0.02 *M* Tris, pH 8.0, 0.02 mol of Tris base is dissolved in 1.0 L of distilled water.

$$0.02 \text{ mol} = 0.02 \times 121 \text{ g} = 2.42 \text{ g of Tris base (solid)}$$

Initially, the pH of this solution will depend to some extent on the pH of the distilled water but will generally be above pH 8.3. To obtain the desired pH 8.0, small amounts of 0.1 *N* HCl are added with stirring until the pH drops to 8.0.

(b) See Problem 2–23 for the dissociation of these buffer components in an aqueous solution. To make this 0.2 *M* "phosphate" buffer, the ratio of $[A^-]/[HA]$ must be determined for pH 7.0. At pH 7.0,

$$7.0 = 6.86 + \log \frac{[A^-]}{[HA]}$$

where $[A^-] = [HPO_4^{2-}]$ and $[HA] = [H_2PO_4^-]$.

$$0.14 = \log \frac{[A^-]}{[HA]}; \quad \text{antilog } 0.14 = 1.38$$

$$1.38 = \frac{[A^-]}{[HA]}$$

Then the ratio of the buffer components, Na_2HPO_4/NaH_2PO_4, must be 1.38, assuming that Na_2HPO_4 provides the bulk of A^- and NaH_2PO_4, the HA. Since 1.0 L of a 0.2 *M* solution is required, then mol A^- + mol HA must equal 0.2 mol and mol A^-/mol HA must equal 1.38. Substituting for A^-, we obtain

$$\frac{0.2 \text{ mol} - \text{mol HA}}{\text{mol HA}} = 1.38$$

$$0.2 - \text{HA} = 1.38\text{HA}$$

$$2.38\text{HA} = 0.2$$

$$\text{HA} = \frac{0.2}{2.38} = 0.084 \text{ mol}$$

$$A^- = 0.2 - 0.084 = 0.116 \text{ mol}$$

Since HA is NaH_2PO_4, MW = 122; thus

$$0.084 \times 122 = 10.24 \text{ g of } NaH_2PO_4 \text{ is required}$$

A^- is Na_2HPO_4, MW = 142; thus

$$0.116 \times 142 = 16.47 \text{ g of } Na_2HPO_4 \text{ is required}$$

So to make this phosphate buffer, 10.24 g of NaH_2PO_4 and 16.47 g of Na_2HPO_4 are added to 1.0 L of distilled water.

Polyprotic Acids

Many small molecules that are essential in metabolic pathways in the cell have more than one ionizable group. These compounds are polyprotic and are capable of releasing two or more protons, as compared to monoprotic compounds such as lactic acid, which has only one dissociable proton (H^+).

Polyprotic acids have *different pK_a's for each dissociable proton.* Although a molecule may contain more than one carboxyl group, each has a different pK_a. Also, the pK_a for any group is characteristic of the molecule in which it resides. For example, the three carboxyl groups of citric acid have pK_a values different from each other and different from the three found in a similar molecule, isocitric acid.

Diprotic Acids

In the preceding section, the compound lactic acid, a monoprotic acid, was used. Its carboxyl group has a pK of 3.86; that is, at pH 3.86, half of the lactate molecules are protonated, the other half are unprotonated.

Malic acid is a diprotic acid, with two dissociable H^+ (protons) and may exist in three forms, depending on the pH.

pK 3.5:   COOH   COO⁻   COO⁻

           | | |

       HC—OH   HC—OH   HC—OH

        | | |

        CH₂   CH₂   CH₂

        | | |

pK 5.0:   COOH   COOH   COO⁻

       (a)   (b)   (c)

net charge =   0   −1   −2

Problem 2-25

Determine the concentration of *fully unprotonated* malic acid (charge = −2) at pH 5.4. The concentration of malate molecules (all forms together) is 0.3 *M*.

Solution:

To determine the amount of malic acid in the fully unprotonated form at pH 5.4, the Henderson–Hasselbalch equation is used. The pK that is used is 5.0, since at pH 5.4, we assume that the COOH with pK 3.5 would be fully *unprotonated.* Then

$$pH = pK + \log \frac{[A^-]}{[HA]}$$

$$5.4 = 5.0 + \log \frac{[A^-]}{[HA]}$$

$$0.4 = \log \frac{[A^-]}{[HA]} \; ; \; \text{antilog } 0.4 = 2.5$$

$$2.5 = \frac{[A^-]}{[HA]}$$

Let $[A^-] + [HA] = 0.3$ M, so $[HA] = 0.3 - [A^-]$. Then substitute:

$$2.5 = \frac{[A^-]}{0.3 - [HA]}$$

$$2.5(0.3 - [A^-]) = [A^-]$$

$$0.75 - 2.5[A^-] = [A^-]$$

$$0.75 = 3.5[A^-]$$

$$\frac{0.75}{3.5} = [A^-] = 0.21 \; M$$

Therefore, the concentration of the fully unprotonated form of malic acid at pH 5.4 is 0.21 M.

Problem 2-26

Calculate the concentration of malic acid molecules with a net charge of -1 in a solution, ph 4.9, where the concentration of all forms of malic acid is 3.1 mM.

Solution:

The only form of malic acid with a net charge of -1 is

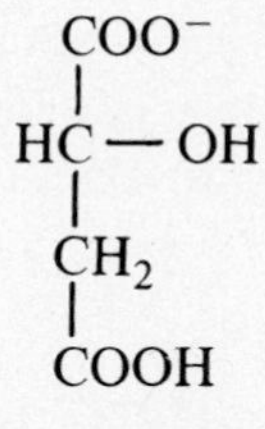

$$\begin{array}{c} COO^- \\ | \\ HC-OH \\ | \\ CH_2 \\ | \\ COOH \end{array}$$

which is the protonated form of the second carboxyl (see p 64). The pK 5.0 is used since at pH 4.9 almost all the carboxyl groups, pK 3.5 would be unprotonated. Using the Henderson-Hasselbalch equation, we obtain

$$pH = pK + \log \frac{[A^-]}{[HA]}$$

$$4.9 = 5.0 + \log \frac{[A^-]}{3.1 - [A^-]}$$

$$-0.1 = \log \frac{[A^-]}{3.1 - [A^-]}$$

$$\text{antilog}\,(-0.1) = 0.794$$

$$0.794 = \frac{[A^-]}{3.1 - [A^-]}$$

$$2.46 = 1.79[A^-]$$

$$[A^-] = 1.37 \text{ m}M$$

Since $[A^-] + [HA] = 3.1$ mM, then $[HA] = 3.1$ m$M - [A^-]$. The concentration of the malic acid species with a net charge of -1 ($[HA]$) is 3.1 m$M - 1.37$ mM or 1.73 mM.

Triprotic Acids

The *tricarboxylic acid cycle* (TCA) is one of the most important metabolic pathways in cells. The cycle is named for its intermediates, citrate, *cis*-aconitate, and isocitrate, each of which has three carboxyl groups. Each carboxyl group may dissociate a proton, and therefore these molecules are termed *triprotic acids*.

The different forms of citrate are shown below with approximate pK_a values for each carboxyl group.

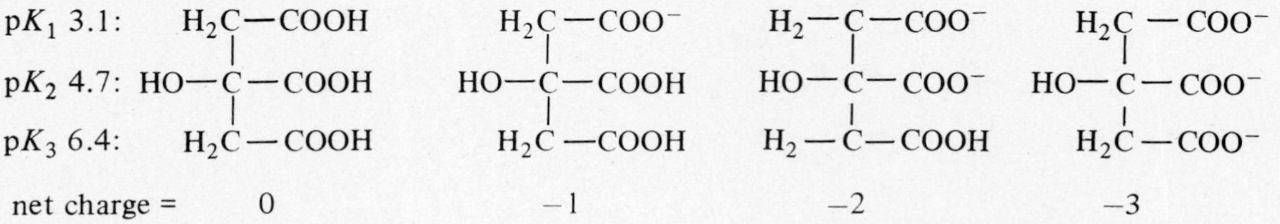

pK_1 3.1:	H$_2$C—COOH	H$_2$C—COO$^-$	H$_2$—C—COO$^-$	H$_2$C —COO$^-$
pK_2 4.7:	HO—C—COOH	HO—C—COOH	HO—C—COO$^-$	HO—C—COO$^-$
pK_3 6.4:	H$_2$C—COOH	H$_2$C—COOH	H$_2$—C—COOH	H$_2$C—COO$^-$
net charge =	0	-1	-2	-3

Problem 2-27

The enzyme aconitase is thought to bind only fully unprotonated citrate molecules (net charge = -3). In a solution at pH 6.9, the concentration of all forms of citrate is 200 μM. What is the concentration of citrate that could bind to the aconitase enzyme? (The pK values for the three functional groups are given above.)

Solution:

Let [unprotonated citrate] = x μM; then [protonated citrate] = $200 - x$. Then the Henderson–Hasselbalch equation may be used to determine [unprotonated citrate], using p$K = 6.4$.

$$6.8 = 6.4 + \log \frac{[A^-]}{200 - [A^-]}$$

$$0.4 = \log \frac{[A^-]}{200 - [A^-]}$$

$$\text{antilog}\,(0.4) = 2.51$$

$$2.51 = \frac{[A^-]}{200 - [A^-]}$$

$$2.51(200 - [A^-]) = [A^-]$$

$$502 - 2.51[A^-] = [A^-]$$

$$502 = 3.51[A^-]$$

$$[A^-] = \frac{502}{3.51} = 143.0 \;\mu M$$

Therefore, the concentration of unprotonated citrate is $143.0 \;\mu M$.

Amino Acids Are Polyprotic Acids

Free amino acids contain at least two groups that can dissociate protons. The α-carboxyl group of free amino acids is a moderately strong acid with pK_a around 2.1 (see Table 2–2). The α-amino group of free amino acids is a weak base with pK_b values ranging from 9.0 to 10.

The general structure for an amino acid at a pH near 7 is

As the pH increases, protons dissociate from the α-amino group,

Several amino acids are triprotic, having three functional groups, each capable of dissociating a proton. The third group is on the R group or side chain of its amino acid. In Table 2–2 the amino acids arg, asp, cys, glu, his, lys, and tyr are triprotic and therefore have pK_a values for α-carboxyl, α-amino, and the R group.

Problem 2–28

Predict the shape of a titration curve for the amino acid alanine, starting with a solution at pH 1.9 and titrating with NaOH. The pK_a values for alanine may be obtained from Table 2–2.

Table 2-2. Structures and pK Values for the 20 Common Amino Acids[a]

Amino Acid	Formula (fully Protonated)	$pK_{1\text{-}COOH}$	$pK_{\alpha\text{-}NH_3}$	$pK_{R\ group}$
Alanine (ala)	$H_3C-\overset{\overset{+}{N}H_3}{\underset{H}{C}}-COOH$	2.34	9.69	—
Arginine (arg)	$H_2N-\overset{\overset{+}{N}H_3}{C}-N-CH_2-CH_2-CH_2-\overset{\overset{+}{N}H_3}{\underset{H}{C}}-COOH$	2.17	9.04	12.48 $\left(H_2N-\overset{\overset{+}{N}H_2}{C}-N-\underset{H}{}\right)$
Asparagine (asn)	$H_2N-\overset{O}{C}-CH_2-\overset{\overset{+}{N}H_3}{\underset{H}{C}}-COOH$	2.02	8.80	—
Aspartic acid (asp)	$HOOC-CH_2(\beta)-\overset{\overset{+}{N}H_3}{\underset{H}{C}}-COOH$	1.88	9.60	3.85 (β-COOH)
Cysteine (cys)	$HS-CH_2-\overset{\overset{+}{N}H_3}{\underset{H}{C}}-COOH$	1.96	10.28	8.18 $\left(HS-\underset{|}{\overset{|}{C}}-\right)$
Glutamic acid (glu)	$HOOC(\gamma)-CH_2-CH_2-\overset{\overset{+}{N}H_3}{\underset{H}{C}}-COOH$	2.19	9.67	4.25 (γ-COOH)
Glutamine (gln)	$H_2N-\overset{O}{C}-CH_2-CH_2-\overset{\overset{+}{N}H_3}{\underset{H}{C}}-COOH$	2.17	9.13	—
Glycine (gly)	$H-\overset{(\alpha)\overset{+}{N}H_3}{\underset{H}{C}}-COOH\ (1)$	2.34	9.60	—
Histidine (his)	$\overset{}{\underset{HN\ \ \ NH}{\diagup}}-CH_2-\overset{\overset{+}{N}H_3}{\underset{H}{C}}-COOH$	1.82	9.17	6.00 $\left(\underset{HN\ \ NH}{\diagup}\right)$

[a]Amino acids exist in solution as zwitterions (i.e., as dipolar ions with both positive and negative charges). In many textbooks, the uncharged form of amino acids is used, but this form is very rarely seen in nature. It is important to see the difference between the uncharged form and the dipolar form, which often has a net charge of zero.

$$R-CH\diagup^{NH_3{}^+}_{\diagdown COO^-} \qquad R-CH\diagup^{NH_2}_{\diagdown COOH}$$

dipolar form of generalized amino acid (net charge 0)   uncharged form

Table 2-2 (continued)

Amino Acid	Formula (Fully Protonated)	pK_{1-COOH}	$pK'_{\alpha-NH_3^+}$	$pK'_{R\ group}$
Isoleucine (ile)	H_3C-CH_2 … $H-C-C-COOH$ with NH_3, H_3C, H	2.36	9.60	—
Leucine (leu)	H_3C, H_3C … $H-C-CH_2-C-COOH$ with $\overset{+}{N}H_3$, H	2.36	9.60	—
Lysine (lys)	(ϵ) $\overset{+}{H}N(H)-CH_2-CH_2-CH_2-CH_2-C-COOH$ with $\overset{+}{N}H_3$, H	2.18	8.95	10.53 (ϵ-NH$_3$)
Methionine (met)	$H_3C-S-CH_2-CH_2-C-COOH$ with $\overset{+}{N}H_3$, H	2.28	9.21	—
Phenylalanine (phe)	$C_6H_5-CH_2-C-COOH$ with $\overset{+}{N}H_3$, H	1.83	9.13	—
Proline (pro)	pyrrolidine ring with $\overset{+}{N}H$ and COOH	1.99	10.60	—
Serine (ser)	$HOC(H)(H)-C-COOH$ with $\overset{+}{N}H_3$, H	2.21	9.15	—
Threonine (thr)	$HO-C(H)(CH_3)-C-COOH$ with $\overset{+}{N}H_3$, H	2.09	9.10	—
Tryptophan (trp)	indole-$CH_2-C-COOH$ with $\overset{+}{N}H_3$, H	2.83	9.39	—
Tyrosine (tyr)	$HO-C_6H_4-CH_2-C-COOH$ with $\overset{+}{N}H_3$, H	2.20	9.11	10.07 ($HO-C_6H_4-$)
Valine (val)	H_3C, H_3C … $H-C-C-COOH$ with $\overset{+}{N}H_3$, H	2.32	9.62	—

Solution:

From Table 2–2 the pK values for alanine are 2.34 and 9.69. The pH at 0.5 equivalent = pK_a for the carboxyl group (2.34). The pH at 1.5 equivalents = 9.69. The pH at 1.0 equivalent = 2.34 + 9.69/2 = 5.96. A smooth curve may be drawn to indicate the plateaus of buffering around the pH = pK_a for each group, as shown in Fig. 2–3.

Problem 2–29

Show structures for the three forms in which alanine may exist, the net charge on each, and the approximate pH ranges where each form predominates. Table 2–2 gives the alanine structure and pK_a values.

Solution:

The three forms of alanine are

$$\overset{+}{N}H_3-CH-COOH \qquad \overset{+}{N}H_3-CH-COO^- \qquad \overset{+}{N}H_3-CH-COO^-$$
$$\quad\;\; | \qquad\qquad\qquad\quad | \qquad\qquad\qquad\quad |$$
$$\quad CH_3 \qquad\qquad\qquad CH_3 \qquad\qquad\qquad CH_3$$

Net charge: +1   0   −1

Approximate
pH range: 0–2.34   2.34–9.69   9.69–14

Problem 2–30

Calculate the concentration of serine molecules with a net charge of +1 in a 20 mM serine solution at pH 3.0.

Solution:

From Table 2–2 the pK_a values for glycine are 2.21 and 9.15. At pH values near 3.0, the forms of glycine that predominate are:

$$\overset{+}{N}H_3-CH-COOH \qquad\qquad \overset{+}{N}H_3-CH-COO^-$$
$$\qquad\quad | \qquad\qquad\qquad\qquad\qquad | $$
$$\qquad\quad H \qquad\qquad\qquad\qquad\qquad\; H$$

Net charge: +1   0

The pK_a of the carboxyl group involved is 2.21. Since the pH is more than 2 pH units below the pK for the amino group, we assume that all the amino groups will remain protonated.

Let HA represent serine with a net charge of +1 and A$^-$ the serine molecules with a net charge of 0. Since [HA] + [A$^-$] = 20 mM, then [A$^-$] = 20 mM − [HA]. Substituting in the Henderson–Hasselbalch equation, we obtain

$$pH = pK + \log \frac{[A^-]}{[HA]}$$

$$3.0 = 2.21 + \log \frac{20 - [HA]}{[HA]}$$

$$0.79 = \log \frac{20 - [HA]}{[HA]}; \quad \text{antilog } (0.79) = 1.2$$

$$1.2 = \frac{20 - [HA]}{[HA]}$$

$$2.2[HA] = 20$$

$$[HA] = 9.09 \text{ m}M$$

Problem 2–31

Some textbooks show the structure for amino acids in an uncharged form:

$$R - \underset{\underset{COOH}{|}}{CH} - NH_2$$

At what pH would this form of amino acid predominate assuming $pK_{a1} = 2.5$ and $pK_{a2} = 9.5$?

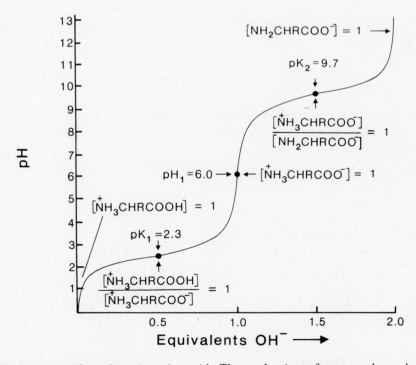

Figure 2–3. Titration curve for a diprotic amino acid. The predominate forms are shown below and are indicated on the titration curve.

Solution:

Inspection of the titration curve for this amino acid (Fig. 2–3) reveals that the uncharged form does not predominate at any pH. In fact, the uncharged form occurs with a frequency of less than 1 per million molecules.

Problem 2–32

An amino acid binding protein in cell membranes has been shown to bind histidine, but only histidine molecules, with a net charge of +1. Calculate [his⁺] in 20 μM histidine solution at pH 6.4, assuming that the pK_a's for histidine are 1.82, 6.00, and 9.17 (Table 2–2).

Solution:

Since the pH is more than 2 full pH units above the pK_a for the carboxyl group, we assume that virtually all carboxyl groups are unprotonated. The pH is more than 2 pH units below the pK_a for the amino group, so we assume that all amino groups are protonated.

Therefore, at pH 6.4, the predominate forms of histidine will be

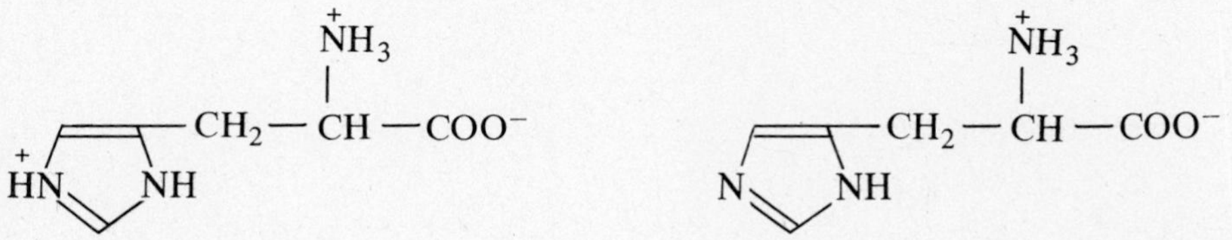

Using the Henderson–Hasselbalch equation to calculate [his⁺], we are concerned with the dissociation of protons from the imidazole group, pK_a 6.00. Then his⁺ is the weak acid HA and his⁰ its conjugate base A⁻.

Since [his⁰] + [his⁺] = 20 μM, [his⁰] = 20 μM − [his⁺]. Substituting yields

$$pH = pK_a + \log \frac{[his^0]}{[his^+]}$$

$$6.4 = 6.0 + \log \frac{20 - [HA]}{[HA]}$$

$$0.4 = \log \frac{20 - [HA]}{[HA]} \; ; \quad \text{antilog} \, (0.4) = 2.5$$

$$2.5 = \frac{20 - [HA]}{[HA]}$$

$$2.5[HA] = 20 - [HA]$$

$$3.5[HA] = 20$$

$$[HA] = \frac{20}{3.5} = 5.71$$

The concentration of his with a net charge of +1 is 5.71 μM.

Peptides as Proton Donors and Acceptors

Peptides are linear polymers of amino acids linked by peptide bonds. Peptide bonds are formed by eliminating H_2O, the OH from the COOH group of one amino acid, and a H atom from the amino group of another amino acid. The ionizable R groups are not involved in forming the peptide bonds.

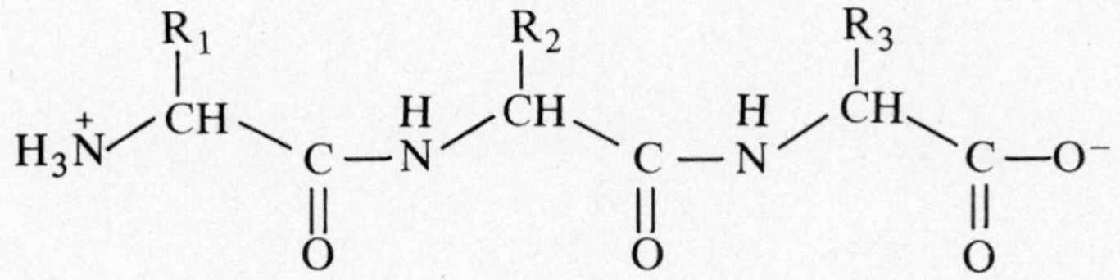

Only the α-amino group at the N terminal in the peptide, the α-carboxyl group at the C terminal, and the ionizable R groups remain capable of dissociating protons. The pK_a values for titratable groups are significantly different in the peptide than in the corresponding free amino acid (Table 2–3). Each pK_a value is influenced by its environment in the peptide, and no single value is correct for the same titratable group. The dissociation of protons from the phenolic OH on tyr and the $-$SH on cys are less predictable in peptides.

Problem 2–33

Calculate the net charge on the peptide, gly-leu-ala, at pH values of 1.0, 3.3, 6.8, and 11.5.

Solution:

Write the structure of the peptide at pH 7 indicating the pK_a for each titratable group. Assume that the pK_a values are the midpoints for the ranges given in Table 2–3. At pH 7.0:

Table 2-3. Approximate pK$_a$ for Titratable Groups in Peptides

Group	pK$_a$ range in proteins
$-NH_3^+$ (N terminal)	7.5–8.4
$-COOH$ (C terminal)	3.0–3.6
$-COOH$ (asp)	3.0–3.5
$-COOH$ (glu)	4.2–4.7
$-$Imidazole (his)	5.6–7.0
$-NH_3^+$ (lys)	9.4–10.6
$-NH-C-NH_2$ (arg) $\quad\quad\;\; |$ $\quad\quad\; NH$	12.0–12.9

$$\overset{+}{NH_3}-CH_2-\underset{\underset{O}{\|}}{C}-\underset{\underset{H}{|}}{N}-\underset{\underset{CH_2}{|}}{CH}-\underset{\underset{O}{\|}}{C}-\underset{\underset{H}{|}}{N}-\underset{\underset{CH_3}{|}}{CH}-O-O^-$$

$$CH-CH_3$$
$$|$$
$$CH_3$$

pK$_a$:   7.95   3.3

Determine the net charge at the different pH values, by adding the charges on each titratable group.

$$\text{N terminal} \quad\quad \text{C terminal}$$

At pH 1.0:   net charge =   +1   +   0   =   +1

At pH 3.3:   net charge =   +1   + $\frac{1}{2}(-1)$ = $-\frac{1}{2}$

Since pH = pK$_a$, one half of the peptide molecules will have their C terminals protonated and the other half unprotonated. Therefore, the net charge for the C terminal $= \frac{1}{2}(0) + \frac{1}{2}(-1) = -\frac{1}{2}$.

$$\text{N terminal} \quad\quad \text{C terminal}$$

At pH 6.8:   net charge =   +1   +   (−1)   =   0

At pH 11.5   net charge =   0   +   (−1)   =   −1

Problem 2-34

Calculate the net charge on the following peptide at pH 1.0, 6.3, and 10.1.

met-his-arg-phe-glu

N terminal   C terminal

Solution

Write the structure of the peptide at pH 7.0, indicating the pK_a for each titratable group.

$$\overset{+}{N}H_3-CH-\overset{\overset{\displaystyle H}{|}}{C}-N-CH-\overset{\overset{\displaystyle O}{\|}}{C}-N-CH-\overset{\overset{\displaystyle O}{\|}}{C}-N-CH-\overset{\overset{\displaystyle O}{\|}}{C}-N-CH-COO^-$$

| | | | | | | | | |
| R_{met} | O | R_{his} | | R_{arg} | | R_{phe} | | R_{glu} |

pK_a: 7.95   6.3   12.45   4.25   3.3

Determine the net charge by adding the charges on each titratable group at each different pH.

At pH 1.0:

| | | | | | | Net charge |
| +1 | +1 | +1 | +0 | +0 | = | +3 |

At pH 6.3:

| +1 | $+\frac{1}{2}$ | +1 | -1 | -1 | = | $+\frac{1}{2}$ |

At pH 10.1:

| +0 | +0 | +1 | -1 | -1 | = | -1 |

Problem 2-35

Show the titration curve for the peptide lys-pro-his, titrating with HCl and starting with the fully unprotonated form, at pH 11.0.

Solution:

The titratable groups are:

N-terminal NH_3 (lys), pK_a 7.95

Secondary NH_3 (lys), pK_a 10.0

Imidazole NH (his), pK_a 6.3

C-Terminal COOH (his), pK_a 3.3

Plot the pK_a values. Since the titration begins at pH 11.0, $x_0 = 0, y_0 = 11.0$:

$x_1 = 0.5$ equivalent, $y_1 = 10.5$

$x_2 = 1.5$ equivalent, $y_2 = 7.95$

$x_3 = 2.5$ equivalent, $y_3 = 6.3$

$x_4 = 3.5$ equivalent, $y_4 = 3.3$

Plot the inflection points.

$$x_5 = 1.0, y_5 = \frac{10.0 + 7.95}{2} = 8.97$$

$$x_6 = 2.0, y_6 = \frac{7.95 + 6.3}{2} = 7.13$$

$$x_7 = 3.0, y_7 = \frac{6.3 + 3.3}{2} = 4.8$$

Draw a smooth curve indicating the areas of buffering around the pK_a's (Fig. 2–4).

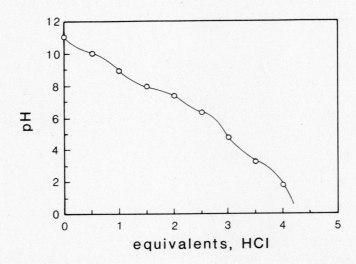

Figure 2-4. Titration curve for lys-pro-his.

pI Determinations

The pI is the isoelectric point, which is defined as the pH at which the net charge on the molecules in solution is zero. When the solution contains a *diprotic acid* only, such as glycine, the pI is simply the average of the two pK values. For example, the pK values for the titratable groups on glycine are 2.3 and 9.6. $pI = (2.3 + 9.6)/2 = 5.95$.

Problem 2-36

Calculate pI for each compound.

(a) Phenylalanine, pK's = 2.2 and 9.5

(b) Methionine, pK's = 2.2 and 9.6

Solution:

(a) $pI = \dfrac{2.2 + 9.5}{2} = 5.85$

(b) $pI = \dfrac{2.2 + 9.6}{2} = 5.90$

Calculating pI when there are more than two ionizable groups is similar in that the pI is an average of two pK_a's, but first we must determine which two pK's to use. In finding the pI for the amino acid, histidine, we must write all the possible charged forms at the different pH's.

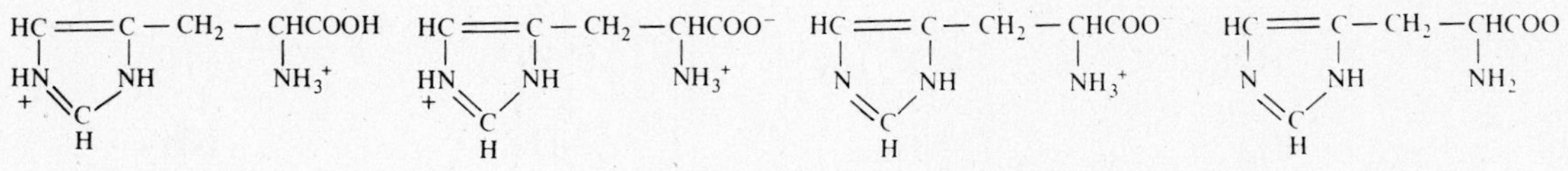

| +2 | +1 | 0 | 1 |

Since pI is defined as the pH at which the net charge is zero on all molecules, we must identify the two pK's between which that condition will occur. By inspection of the forms of histidine above, it is clear that we are seeking the pH at which all molecules have lost the protons from the COOH group and the R (imidazole) group, but not the α-NH_3 group. That pH is the average between the pK for the R group (6.0) and the pK for the $-NH_3$ group (9.2).

$$pI = \frac{6.0 + 9.2}{2} = 7.6$$

Problem 2–37

Calculate pI for the following molecules using the pK values given in Table 2–2 for free amino acids and in Table 2–3 for the peptide.

(a) Glutamic acid

(b) Lysine

(c) The peptide his-ala-thr

Solution:

(a)

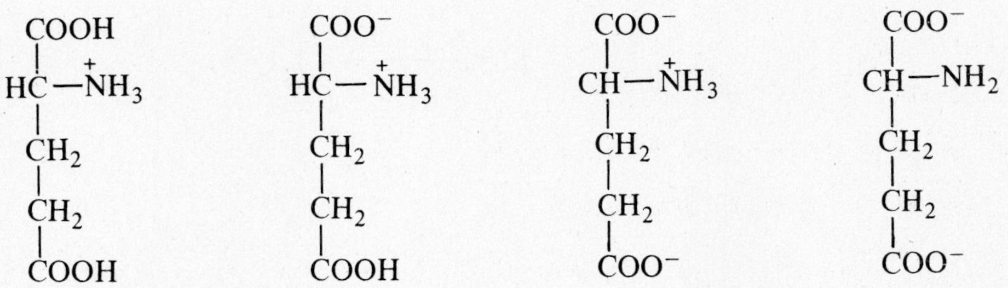

| Net change: | +1 | 0 | −1 | −2 |

After inspecting the different structures for glu at the various pH's, the pI must lie between the pK for the primary COOH (2.2) and the pK for the secondary COOH (4.2):

$$pI = \frac{2.2 + 4.2}{2} = 3.2$$

(b)

$$
\begin{array}{cccc}
\text{COOH} & \text{COO}^- & \text{COO}^- & \text{COO}^- \\
| & | & | & | \\
\text{HC}-\overset{+}{\text{NH}_3} & \text{HC}-\overset{+}{\text{NH}_3} & \text{HC}-\text{NH}_2 & \text{HC}-\text{NH}_2 \\
| & | & | & | \\
(\text{CH}_2)_4 & (\text{CH}_2)_4 & (\text{CH}_2)_4 & (\text{CH}_2)_4 \\
| & | & | & | \\
\underset{+}{\text{NH}_3} & \underset{+}{\text{NH}_3} & \underset{+}{\text{NH}_3} & \text{NH}_2
\end{array}
$$

Net charge: +2   +1   0   −1

From the different structures for lysine, the pI must lie between the pK for the primary $-\text{NH}_3$ and the secondary $-\text{NH}_3$.

$$pI = \frac{8.9 + 10.5}{2} = 9.7$$

(c) By convention, the N terminal of the peptide is to the left and the C terminal to the right.

$$
\text{H}_3\overset{+}{\text{N}}-\underset{\underset{\text{R}_{his}}{|}}{\text{CH}}-\overset{\overset{\text{O}}{\|}}{\text{C}}-\underset{\underset{\text{H}}{|}}{\text{N}}-\underset{\underset{\text{R}_{ala}}{|}}{\text{CH}}-\overset{\overset{\text{O}}{\|}}{\text{C}}-\underset{\underset{\text{H}}{|}}{\text{N}}-\overset{\overset{\text{R}_{thr}}{|}}{\underset{\underset{\text{COO}^-}{|}}{\text{HC}}}
$$

pK: 7.9   6.3   3.3

The peptide his-ala-thr has three titratable groups with pK's at 7.9 for the primary amino group of his, 6.3 for the imidazole group, and 3.3 for the COO$^-$ group of thr. From inspecting the different possible charged forms of this peptide, the pI must lie between the amino and the imidazole groups of his.

$$pI = \frac{6.3 + 7.9}{2} = 7.1$$

Exercises

The answers are provided at the back of the book.

2-1. The dissociation constant, K_a for propionic acid is 1.35×10^{-5}. Calculate [propionic acid] in a solution containing 100 μM propionate ions and 50 μM H^+ ions.

2-2. What is the K_a for the carboxyl group of histidine given that pK_a is 1.82?

2-3. What is [lactate]/[lactic acid] in a 90 μM solution of lactic + lactate at pH 5.8? The pK_a for lactic acid is 3.87.

2-4. What is [acetate ion] in a 0.1 M acetic acid solution, pH 5.0, assuming that the pK_a for acetic acid is 4.7.

2-5. What is [HCO_3^-] in an aqueous solution of 0.01 M $NaHCO_3$, pH 6.0? The pK_a for $H_2CO_3 \rightleftharpoons HCO_3^- + H^+$ is 6.1.

2-6. Determine the concentration of histidine molecules with a net charge of 0 in a 10 mM histidine solution at pH 6.8.

2-7. Determine the net charge on the peptide met-ala-thr-asp-arg-pro-leu at pH 5.1.

2-8. Calculate pI for:

(a) Glutamine

(b) The peptide in Exercise 2-7

2-9. At which points on the titration curve for glycine would the following occur?

(a) [gly^+] = [gly^0]

(b) [gly^-] = [gly^0]

(c) All gly has a net charge of zero

gly^+ represents glycine with a net charge of +1; gly^- = glycine with a net charge of −1, and gly^0 = glycine with a net charge of zero.

2-10. How much solid Na acetate and acetic acid is required to form 500 mL of 0.03 M acetate buffer, pH 5.2? The MW of sodium acetate is 72, and of acetic acid is 60. The pK_a for acetic acid is 4.87.

Chapter 3

ENZYME KINETICS

Enzymes are proteins that act as catalysts in cells, increasing the rates of chemical reactions. In the absence of enzymes, chemical reactions would still occur, but at rates so low as not to support life. In the presence of an enzyme the rate of a reaction may increase by 10,000 or even 10,000,000 times. Consider Fig. 3-1, where substrate A $\longrightarrow$ product B in the presence and absence of catalysts.

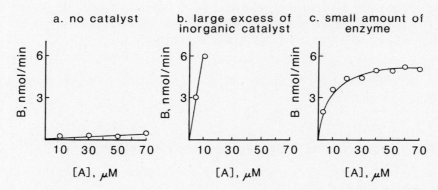

Figure 3-1. Rates of product formation at increasing concentrations of substrate A in the presence and absence of catalysts.

In Fig. 3-1a, no catalyst is present and very little, if any, product B is formed during the period in which the reaction occurs (e.g., 1 min). In Fig. 3-1b an inorganic catalyst *in large excess* is present and the amount of product B formed in the same period increases enormously. In fact, so much product is formed that the values obtained at higher concentrations of A are off the graph. In Fig. 3-1c there is a small amount of enzyme acting as catalyst and the rate of product formed is similar to Fig. 3-1b using the inorganic catalyst, but only at lower concentrations of substrate A. *In cells, enzymes are present in very small amounts* and *act to increase the rates of chemical reactions,* generally at lower substrate concentrations. As the concentration of substrate A increases, the enzyme catalyst becomes saturated and the amount of enzyme is the factor limiting the rate of the reaction.

Enzymes bind substrates to a specific region called the *active site*. This region of the protein contains amino acids such as histidine, lysine, tyrosine, aspartate, cysteine, and serine.

The substrate is bound by electrostatic attractions, hydrogen bonding, or by hydrophobic interactions to the R groups of the amino acids at the active site. Most enzymes bind only a few structurally related molecules to their active sites. If the active site of an enzyme is changed, this usually decreases the effectiveness in catalyzing the reaction and often makes the enzyme totally inactive.

Measuring Rates of Enzyme-Catalyzed Reactions

Enzyme kinetics is the study of rates of enzyme catalysis. Almost all enzyme-catalyzed reactions are studied outside the cell, in test tubes (i.e., *in vitro*). Many different enzymes have been purified from extracts of the cells in which they were formed. Often, nanogram quantities of an enzyme (10^{-9} g) are all that is required to produce μmoles of product in a 1.0-min period *in vitro*.

It is important to realize that in cells (i.e. in vivo) the substrate concentrations are much, much larger than enzyme concentrations, but generally the SUBSTRATE CONCENTRATION, not the enzyme concentration IS THE LIMITING FACTOR IN THE REACTION. In most cases, the substrate concentrations are close to, or slightly smaller than, the value of K_m, the Michaelis constant (derived below).

Rates of enzyme catalysis are measured as the amount of product formed per time unit and are usually expressed as *μmoles of product formed per minute per mg of enzyme*. This is called the *velocity* of the reaction. Michaelis and Menten developed a model to describe enzyme-catalyzed reactions mathematically based on certain specific assumptions. Using this model it is possible to determine the *maximum initial velocity* ($V_{\max}$) and the apparent *affinity of the enzyme for its substrate* (K_m). This model is valid *only* if the assumptions made by Michaelis and Menten are true. These assumptions are:

1. An enzyme–substrate complex must be formed as an intermediate to product formation, that is,

   $$E + S \rightleftharpoons ES \longrightarrow E + P$$

   where E is the enzyme, S the substrate, and P the product.

2. The rate of conversion of ES to E + P is significantly slower than ES dissociation and reassociation.

3. Only one S molecule binds to one E molecule:

   $$E + S \rightleftharpoons ES$$

Briggs and Haldane further refined this model by making a fourth assumption:

4. The [S] is very much larger than the [E], so that the [ES] is always constant, at least for the first few minutes of the reaction (the *steady-state* condition). Thus the rate of reaction depends on [ES].

This fourth assumption is true only during the *initial stages* of an enzyme-catalyzed reaction. As S is converted to P, the [S] drops and assumption 4 may not be true several minutes after the reaction has started. Also, if the product is allowed to accumulate, it may inhibit the reaction.

Derivation of the Michaelis–Menten Equation

If we assume that

$$[E] + [S] \underset{k_2}{\overset{k_1}{\rightleftharpoons}} [ES] \overset{k_3}{\longrightarrow} [E] + [P] \tag{1}$$

where k_1, k_2, and k_3 are the rate constants, the overall reaction rate (μmol P formed/min) is dependent on [ES]. Since we assume that [ES] is constant, the rate of ES formation must equal the rate of ES disappearance and may be expressed as:

$$
\begin{array}{cc}
\textit{rate of ES formation} & \textit{rate of ES disappearance} \\
k_1[E][S] \qquad = & k_2[ES] + k_3[ES]
\end{array} \tag{2}
$$

where [E] represents free enzyme concentration.

This equation may be rearranged to give

$$[ES] = \frac{[E_t][S]}{[S] + k_2 + k_3)/k_1} \tag{3}$$

where $[E_t]$ = total E or [ES] + [E].

The term $(k_2 + k_3)/k_1$ is defined as the *Michaelis constant, K_m*. If we assume that k_2 (rate constant for ES breakdown) is much, much larger than k_3 (rate constant for P formation), then K_m approximates k_2/k_1, which is K_s, the dissociation constant for the ES complex. If all the assumptions above hold (as they do for many enzymes), K_m *may be used as a measure of the affinity of an enzyme for a substrate.* Equation (3) may now be written

$$[ES] = \frac{[E_t][S]}{[S] + K_m} \tag{4}$$

Since the rate-limiting step by assumption 2 is ES $\overset{k_3}{\longrightarrow}$ P, the overall rate of P formation is written

$$v_0 = k_3[ES] \qquad \text{or} \qquad v_0 = k_3[E_t] \tag{5}$$

where v_0 *is the initial velocity, which is the velocity during the initial stages of the reaction* when [S] has not decreased significantly or very little product has formed.

If we combine equations (3) and (4), we obtain

$$v_0 = \frac{k_3[E_t][S]}{[S] + K_m} \tag{6}$$

If we define V_{max} *as the maximum initial velocity* for a given [E] at saturating levels of S, then

$$V_{max} = k_3[ES] \quad \text{or} \quad V_{max} = k_3[E_t] \tag{7}$$

Using this definition, equation (5) becomes

$$v_0 = \frac{V_{max}[S]}{K_m + [S]} \quad \text{Michaelis–Menten equation} \tag{8}$$

Thus if the K_m and V_{max} for a given enzyme have been determined and we know the [S], we may calculate the rate at which product will be formed. Being able to measure the rate of enzyme catalysis allows us to compare enzymes from different sources as to their efficiencies (V_{max}/K_m). Many human disorders are due to defective enzymes, and in many instances, once the lack of enzyme activity has been confirmed, effective treatment is available.

Note:

1. K_m *is the* [S] *at* $v_0 = \frac{1}{2} V_{max}$ and therefore has dimensions of concentration. K_m values for most enzymes lie between 1 μM and 10 mM.
2. V_{max} *is the maximum initial velocity* and has dimensions of μmol P/min/mg enzyme or μmol P/min/mol enzyme.

Problem 3–1

An enzyme catalyzes the reaction A$\longrightarrow$B + H_2O. If K_m for A is 100 μM and V_{max} is 8 μmol/min/mg enzyme, calculate the amount of B formed in 2.0 min by 1.0 mg of enzyme with 50 μM A in solution.

Solution

Using the Michaelis–Menten equation, v_0 (the initial velocity) may be determined. Since v_0 *gives the rate of product formation,* substituting the values given should allow calculation of amount of B formed in 2.0 min by 1.0 mg of enzyme.

$$v_0 = \frac{V_{max}[S]}{K_m + [S]}$$

$$= \frac{(8 \; \mu mol/min)(5 \times 10^{-5} \; M)}{10^{-4} \; M + (5 \times 10^{-5} \; M)} = \frac{(8 \times 10^{-6} \; mol/min)(50 \times 10^{-6} \; mol/L)}{(100 \times 10^{-6} \; mol/L) + (50 \times 10^{-6} \; mol/L)}$$

$$= \frac{400 \times 10^{-12}}{150 \times 10^{-6}} \; mol/min = 2.67 \times 10^{-6} \; mol/min$$

$$= 2.67 \; \mu mol/min/mg \; enzyme$$

Then 1.0 mg of enzyme would catalyze formation of 2.67 μmol of B in 1.0 min and $2 \times 2.67 = 5.34$ μmol of B in 2.0 min.

Problem 3–2

The reaction $A \rightarrow B + H_2O$ is catalyzed by a dehydratase enzyme. A absorbs light strongly at a wavelength of 440 nm. The product, B, however, does not absorb light at any wavelength convenient for measuring with a simple spectrophotometer. The decrease in substrate A was followed spectrophotometrically during the reaction with 0.1 mg of the enzyme in 1.0 mL total volume and the following data were obtained:

At time 0 min: [A] = 0.10 μmol/1.0 mL.

At time 1.0 min: [A] = 0.09 μmol/1.0 mL.

Assuming that this enzyme shows Michaelis–Menten kinetics, calculate the K_m for substrate A if the V_{max} is 0.5 μmol/min/mg dehydratase enzyme.

Solution

Since v_0 is defined as μmol product/min/mg enzyme, the μmol B formed per minute must be determined. According to the reaction stoichiometry, the μmol B formed exactly equals the μmol A consumed. Since

$$\mu\text{mol A consumed} = 0.10 - 0.09 = 0.01 \ \mu\text{mol in 1.0 min}$$

$$v_0 = 0.01 \ \mu\text{mol/min/0.1 mg enzyme}$$

$$= 0.1 \ \mu\text{mol/min/mg}$$

Substituting in the Michaelis–Menten equation, [S] = [A] at time 0 = 100 μmol/L = 100 μM.

Then

$$v_0 = \frac{V_{max}[S]}{K_m + [S]}$$

$$0.1 = \frac{0.5 \times 100}{K_m + 100}$$

$$0.1 K_m + 10 = 0.5 \times 100$$

$$0.1 K_m = 50 - 10$$

$$K_m = \frac{40}{0.1} = 400 \ \mu M$$

Problem 3–3

Lactase, an enzyme found in the small intestine of most human beings, cleaves the milk sugar lactose to yield glucose and galactose. Assuming that K_m for lactose is 0.9 mM, V_{max} is 3.5 μmol/min/mg lactase, 5.7 μmol lactose is present, and the reaction mixture volume is 80 mL, calculate how much glucose would be formed in 1 min by 1 mg of lactase.

Solution:

The number of μmol substrate (lactose) is known, but in the Michaelis–Menten equation, the *concentration* of the substrate is required. To convert from μmol to μM (μmol/L), the number of μmol lactose is divided by the volume of the reaction mixture:

$$[S] = \frac{5.7 \ \mu\text{mol lactose}}{0.080 \ \text{L}} = \frac{71.25 \ \mu\text{mol}}{\text{L}} = 71.25 \ \mu M$$

Since the K_m is given as mM, it should be converted to μM so that the K_m and the [S] will have the same concentration units.

$$K_m = 0.9 \ \text{m}M$$

$$= \frac{0.9 \ \text{mmol}}{\text{L}} \times \frac{1 \ \text{mol}}{10^3 \ \text{mmol}} \times \frac{10^6 \ \mu\text{mol}}{1 \ \text{mol}}$$

$$= 900 \ \mu M$$

$$v_0 = \frac{V_{\text{max}}[S]}{K_m + [S]}$$

$$= \frac{3.5 \ \mu\text{mol/min/mg} \ (71.25 \ \mu M)}{900 \ \mu M + 71.25 \ \mu M}$$

$$= 0.26 \ \mu\text{mol/min/mg}$$

Therefore, 0.26 μmol of glucose will be formed each minute for each milligram of lactase present during the first few minutes of the reaction.

Experimental Determination of K_m and V_{max}

In the problem above, we were able to calculate the amount of product formed per minute (v_0) because the V_{max} and K_m values had been determined previously from experiments to measure initial velocity. The Michaelis constant, K_m, and the maximum initial velocity, V_{max}, must be determined *experimentally* for each enzyme. There is no way to predict K_m and V_{max} without measuring *initial velocities at many different substrate concentrations*. The K_m and V_{max} values have been determined in this way for many enzymes.

Measuring Initial Velocities

Initial velocity (v_0) is defined as the velocity (μmol P formed/min) during the *initial stages* of a biochemical reaction, before [S] is significantly decreased (or P builds up) and becomes rate limiting. To determine K_m and V_{max} experimentally, the v_0 must be measured at different [S]. These types of studies are called *time course assays* and are designed as shown below, with constant [E] at various [S]. In the example below, the amount of product formed in each tube was measured at 30-sec intervals and the results are presented in Table 3–1.

For each [S], the amount of product formed at each time unit is plotted as shown in Fig. 3–2. In the figure, each point represents a velocity but only points *on* the *linear portion* of the curves are *initial velocities.* As the amount of substrate decreases and becomes limiting, or the product increases, the conditions assumed by Michaelis and Menten are no longer present.

Table 3-1. Time Course Assay Data

Assay No.	E (mg)	[S] (μM)	μmol Product Formed			
			30 sec	60 sec	90 sec	120 sec
1	0.01	10.0	0.005	0.010	0.015	0.018
2	0.01	11.1	0.0055	0.011	0.016	0.019
3	0.01	12.2	0.006	0.011	0.017	0.021
4	0.01	14.2	0.006	0.012	0.019	0.023
5	0.01	16.6	0.008	0.017	0.025	0.030
6	0.01	20.0	0.007	0.015	0.022	0.026
7	0.01	25.0	0.008	0.016	0.024	0.028
8	0.01	33.3	0.011	0.022	0.033	0.040
9	0.01	50.0	0.014	0.028	—	—
10	0.01	100.0	0.0165	0.033	—	—

Note: The observed μmol P formed in assay 5 is higher than expected. Observed results often differ slightly from expected since enzyme activity is affected by many factors, including slight temperature changes and inconsistencies in pipetting.

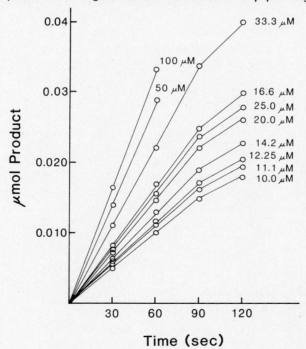

Figure 3-2. Time course assay plots from data in Table 3-1. After 90 sec, the curves become nonlinear, indicating that the [S] has become limiting or the [product] has increased to the point where it inhibits further product formation.

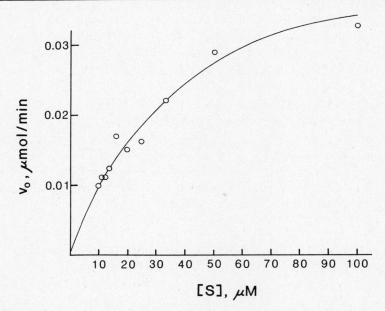

Figure 3-3. Initial velocity (v_0) for each different [S]. Each v_0 is the slope of the linear portion of a time course curve from Fig. 3-2.

By definition, *the slope of the linear portion* = v_0. To obtain v_0 for [S] = 25.0 μM (from Fig. 3–2), we calculate the slope of the *initial linear portion* of the time course curve between 0 and 90 sec. The slope of the line between 0 and 90 sec is 0.016 $\mu mol/60$ sec or 0.016 $\mu mol/min$. The slope after 90 sec is *not an initial velocity* because the line begins to curve as the [S] becomes limiting or the product accumulates. Initial velocities (v_0) for each [S] are shown in Table 3–2. If v_0 is plotted as a function of [S], the graph in Fig. 3–3 is obtained.

Note: v_0 is plotted as μmol product formed/min/0.01 mg enzyme; [S] is in $\mu mol/L$ (μM).

Table 3-2.   Initial Velocities at Different Substrate Concentrations and Their Reciprocal Values

Assay No.	[S] (μM)	v_0 ($\mu mol/min$)	1/[S] (μM^{-1})	$1/v_0$ (min/μmol)
1	10	0.010	0.10	100
2	11.1	0.011	0.09	90.9
3	12.2	0.011	0.08	90.9
4	14.2	0.012	0.07	83.3
5	16.6	0.017	0.63	58.8
6	20.0	0.015	0.05	66.6
7	25.0	0.016	0.04	62.5
8	33.3	0.022	0.03	45.4
9	50.0	0.028	0.02	35.7
10	100.0	0.033	0.01	30.3

Drawing a smooth curve will result in a rectangular hyperbola if all the assumptions for the Michaelis–Menten equation are valid for this enzyme reaction. If, for example, there is more than one substrate-binding site per enzyme molecule, the plot may not resemble a rectangular hyperbola. These situations may be the result of cooperativity and are explored later in this chapter.

V_{max} is defined as the *maximum initial velocity* and is expressed as μmol P formed/min/mg enzyme or μmol P/min/mole enzyme where the molecular weight of the enzyme is known. It is not possible, in most cases, to measure v_0 at a [S] that gives V_{max}. Since it is not possible to extrapolate the value of V_{max} accurately using the Michaelis–Menten plot, this plot is only used to *estimate, never to determine V_{max} accurately*. From Fig. 3-3 we may *estimate* V_{max} to be about 0.035 μmol P formed/min/mg enzyme, but as we shall see below, V_{max} is actually 0.045 μmol/min/mg enzyme.

K_m may be defined as *the [S] when $v_0 = \frac{1}{2} V_{max}$* and has dimensions of concentration. Under certain conditions, K_m may be used as a measure of the affinity of an enzyme for its substrate. A very low K_m value (e.g., 2 μM) indicates a very high affinity for that particular substrate. If it is impossible to measure V_{max} accurately from the Michaelis plot, it follows that the [S] at $v_0 = \frac{1}{2} V_{max}$ cannot be determined either.

Problem 3-4

Explain why in each of the plots shown in Fig. 3-4, the curves become nonlinear.

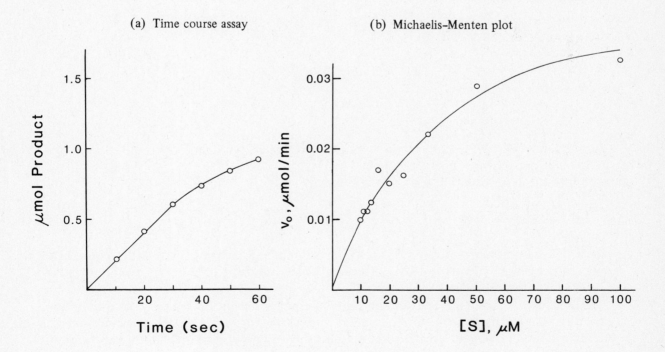

Figure 3-4. (a). Product formation as a function of time, and (b). Rate of product formation at various [S].

Solution:

In Fig. 3–4a, the amount of product formed is plotted at each time unit during the reaction. As substrate is converted into product, *the [S] decreases and becomes limiting,* so that Michaelis–Menten conditions are no longer present and the formation of product is no longer linear with time. In some cases, the product builds up after a few minutes and inhibits the reaction (*product inhibition*), thereby decreasing the amount of product formed per minute.

In Fig. 3–4b, the initial velocity or rate of product formation is plotted for each [S]. At high [S] ([S] = more than $10 \times K_m$), the enzyme is saturated and the amount of enzyme becomes the limiting factor. Even at low [S] this curve is not truly linear, but above [S] = K_m (at $\frac{1}{2} V_{max}$) *the increase in* v_0 *tapers off to almost zero at very high [S], due to saturation of the enzyme.*

Lineweaver–Burk Plot

To allow more accurate determination of K_m and V_{max}, Lineweaver and Burk proposed plotting reciprocal values, which gives a linear relationship. Taking the reciprocals of the Michaelis–Menten equation (8) we obtain

$$\frac{1}{v_0} = \frac{1}{V_{max}[S]/(K_m + [S])} \tag{9}$$

This expression may be rearranged to give

$$\frac{1}{v_0} = \frac{K_m}{V_{max}[S]} + \frac{[S]}{V_{max}[S]} \tag{10}$$

which may be written

$$\frac{1}{v_0} = \left(\frac{K_m}{V_{max}} \times \frac{1}{[S]} \right) + \frac{1}{V_{max}} \tag{11}$$

Note that this is an equation for a straight line,

$$y = (m \times x) + b \tag{12}$$

where $1/v_0$ is represented by y; $1/[S]$ is represented by x; and K_m/V_{max} is the slope (m) of the straight line. The term $1/V_{max}$ (or b) is the y intercept obtained when $1/[S]$ is zero (i.e., [S] is infinitely large).

If the assumptions are valid and the measurements during time course assays are done carefully, *the points approximate a straight line,* as shown in Fig. 3–5. However, since all the data points rarely fall on the same straight line, a *line of best fit is drawn through the points.* From this Lineweaver–Burk plot we can determine V_{max} and K_m much more accurately than from Fig. 3–3. The y intercept gives the reciprocal of V_{max}. K_m may be

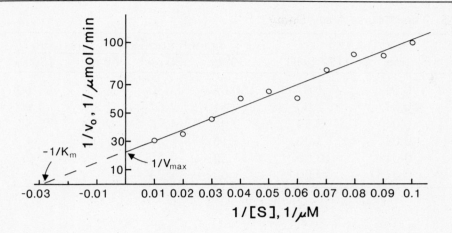

Figure 3-5. Lineweaver–Burk plot. The reciprocal values presented in Table 3-2 are plotted here.

obtained by extrapolating the line back to the x axis (see the dashed line in Fig. 3-5). The negative x intercept equals $-1/K_m$, and K_m has the dimensions of substrate concentration. The K_m and the V_{max} in the example above are determined as:

$$-\frac{1}{K_m} = -0.028 \ \mu M$$

$$K_m = \frac{1}{0.028}$$

$$= 35.7 \ \mu M$$

$$\frac{1}{V_{max}} = 22 \ \mu mol/min/0.01 \ mg \ enzyme$$

$$V_{max} = \frac{1}{22} = 0.045 \ \mu mol/min/0.01 \ mg \ enzyme$$

$$= 4.5 \ \mu mol/min/mg \ enzyme$$

Errors tend to be magnified in the Lineweaver–Burk plot. In research publications, kinetic data are presented in Lineweaver–Burk plots, but the data are treated statistically by nonlinear regression analysis to fit a curve so that K_m and V_{max} may be determined most accurately. In this book the K_m and V_{max} values are approximated using Lineweaver–Burk plots.

Problem 3-5

Determine K_m and V_{max} from the time course assay data (amount of product formed at different [S]) shown in Table 3-3. In each assay, 0.01 mg of an enzyme preparation was used.

Table 3-3. Time Course Assay Data

Tube No.	[E] (mg)	[S] (μM)	nmol P Formed per Time Unit			
			30 sec	60 sec	90 sec	120 sec
1	0.01	1.67	0.65	1.30	1.51	1.60
2	0.01	2.00	0.81	1.62	1.80	1.90
3	0.01	3.33	0.98	1.96	2.30	2.55
4	0.01	5.00	1.16	2.23	3.00	—
5	0.01	10.00	1.24	2.50	—	—
6	0.01	20.00	1.62	3.24	—	—

Solution:

The first step in determining K_m and V_{max} is to plot the amount of product formed as a function of time as shown in Fig. 3-6. From this graph the v_0 may be determined for each [S] by finding the *slope of the linear portion* of each curve. Table 3-4 shows the v_0 values and corresponding reciprocals $(1/[S], 1/v_0)$.

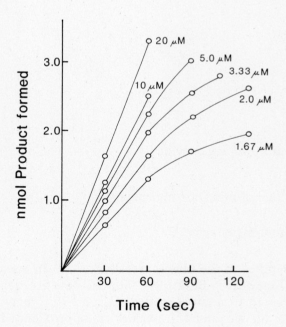

Figure 3-6. Time course assay data from Table 3-3.

Note: The curves for the assays are *not* linear after 60 sec and the slopes must be taken before 60 sec. Velocities measured after 60 sec are *not initial velocities.*

Table 3-4. Initial Velocities (v_0) for Each [S] and Their Reciprocal Values.

[S] (μM)	$v_0{}^a$ (nmol/min/0.01 mg)	1/[S] (μM^{-1})	$1/v_0$ (1/mnol/min/0.01 mg)
1.67	1.30	0.59	0.77
2.00	1.62	0.50	0.62
3.33	1.96	0.30	0.51
5.00	2.32	0.20	0.43
10.00	2.50	0.10	0.40
20.00	3.24	0.05	0.31

$^a\dfrac{n \text{ mol product}}{30 \text{ sec}} \times \dfrac{60 \text{ sec}}{1 \text{ min}} = n \text{ mol product/min}$

To determine K_m and V_{max}, the reciprocals of v_0 and [S] are plotted in the Lineweaver-Burk plot (Fig. 3-7).

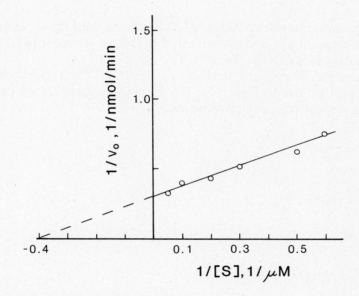

Figure 3-7. Lineweaver–Burk plot of reciprocal values in Table 3-4.

From the double-reciprocal plot the x-intercept, x_0 and the y-intercept, y_0 are -0.4 and 0.3, respectively.

$$x_0 = -0.4 = \frac{1}{K_m}; \quad K_m = 2.5 \ \mu M \qquad x_0 = -0.4 = -\frac{1}{K_m}; \quad K_m = 2.5 \ \mu M$$

$$y_0 = \ 0.3 = \frac{1}{V_{max}} \qquad V_{max} = 3.33 \text{ nmol/min/0.01 mg enzyme}$$

Since V_{max} is usually expressed as μmol/min/mg enzyme:

$$V_{max} = 333 \text{ nmol/min/mg enzyme} \quad \text{or} \quad 0.33 \ \mu\text{mol/min/1.0 mg enzyme}$$

Problem 3-6

Enolase is an enzyme found in glycolysis and catalyzes the formation of phosphoenol-pyruvate from 2-phosphoglycerate.   This enzyme has been isolated and purified.   Time course assays were performed using purified enzyme and varying concentrations of the substrate.   From the data in Table 3-5, determine the V_{max} and the K_m.

Table 3-5. Time Course Assay Data

Assay No.	enzyme (mg)	[S] (mM)	nmol Product Formed		
			1.0 min	2.0 min	3.0 min
1	0.02	0.12	2.3	4.6	6.1
2	0.02	0.16	2.8	5.7	8.2
3	0.02	0.25	3.4	6.8	9.0
4	0.02	0.50	4.0	8.0	11.6
5	0.02	1.00	5.0	9.9	14.6

Solution:

Plot the time course data as shown in Fig. 3-2. From this plot the v_0 values are determined by calculating the slope of the linear portion of the curve for each [S]. Make a table with the v_0 values for each [S] similar to Table 3-4.

Reciprocal values are determined for each v_0 and [S]. The reciprocals are plotted in a Lineweaver-Burk plot similar to Fig. 3-5. The K_m and V_{max} values are calculated from the x and y intercepts, respectively, as shown below.

$$x_0 = -\frac{1}{K_m} = -5.5$$

$$K_m = \frac{1}{5.5} = 0.18 \text{ m}M$$

$$= 0.2 \text{ mM}$$

$$y_0 = \frac{1}{V_{max}} = 0.17$$

$$V_{max} = \frac{1}{0.17} = 5.88 \text{ nmol/min/0.02 mg}$$

Note: These values may vary slightly due to slight differences in drawing the line of best fit.

Since the assays were done with 0.02 mg of enzyme, we must divide by 0.02 to express V_{max} as per milligram of enzyme:

$$V_{max} = 5.88 \text{ nmol/min/0.02 mg enzyme}$$

$$= 294 \text{ nmol/min/mg enzyme}$$

Problem 3-7

The enzyme aldolase catalyzes the cleavage of fructose-1, 6-bisphosphate (F-1, 6-bisP) to give dihydroxyacetone phosphate (DHAP) and 3-phosphoglyceraldehyde (G-3-P). This enzyme is found in glycolysis and the following data were obtained from time course assays using aldolase purified from fungal cell extracts.

$$F\text{-}1,6\text{-bisP} + H_2O \xrightarrow{\text{aldolase}} G\text{-}3\text{-P} + DHAP$$

[F-1,6-bisP] (μM)	v_0 (μmol G-3-P formed/min/mg)
40	0.020
50	0.024
67	0.027
100	0.031

Calculate the K_m and V_{max} for aldolase from the data given above.

Solution:

This enzyme has shown Michaelis–Menten kinetics in previous studies. To determine the K_m and V_{max} from these data, the reciprocals of the substrate concentrations, [F-1, 6-bisP], and the initial velocity values, v_0, are calculated and plotted in the Lineweaver–Burk plot.

[S] (μM)	v_0 (μmol/min/mg)	1/[S] (1/μM)	$1/v_0$ (1/μmol/min/mg)
40	0.020	0.025	50
50	0.024	0.020	42
67	0.027	0.015	37
100	0.031	0.010	32

When these reciprocal values are plotted on the Lineweaver–Burk plot, a straight line may be drawn through the points. This line intersects the x-axis at approximately -0.015 and the y-axis at approximately 18.

Note: Slightly different values may be obtained depending on the line of best fit that you drew.

Then

$$-\frac{1}{K_m} = -0.015$$

$$K_m = 66.67 \ \mu M$$

and

$$\frac{1}{V_{max}} = 18$$

$$V_{max} = 0.055 \ \mu mol/min/mg \quad or \quad 55 \ nmol/min/mg$$

Problem 3-8

The enzyme β-galactosidase found in many bacteria cleaves the disaccharide lactose to yield glucose and galactose. The β-galactosidase purified from the bacterium *Escherichia coli* was used in kinetic studies and gave the following results:

lactose + $H_2O \longrightarrow$ glucose + galactose

[lactose] (mM)	v_0 (μmol/min/0.1 mg enzyme)
0.083	1.61
0.100	1.85
0.125	2.00
0.150	2.50
0.200	2.80

Calculate the K_m and the V_{max} from these data.

Solution:

Assuming that this enzyme obeys Michaelis–Menten kinetics, the reciprocals of the [S] and v_0 values should yield a straight line when plotted. From the intercepts of the line, the K_m and V_{max} values may be determined.

[lactose](mM)	v_0 (μmol/min/mg)	1/[lactose] , mM	1/v_0
0.083	1.61	12	0.62
0.100	1.85	10	0.54
0.125	2.00	8	0.48
0.150	2.50	6.67	0.40
0.200	2.80	5	0.35

A straight line drawn through the points intercepts the x-axis at approximately -0.4 and the y-axis at approximately 0.15. Thus

$$-\frac{1}{K_m} = -0.4$$

$$K_m = 0.25 \ mM$$

and

$$V_{max} = 7.14 \ nmol/min/0.1 \ mg \quad or \quad 71.4 \ nmol/min/mg \ enzyme$$

Problem 3–9

The enzyme threonine dehydratase is found in many cells and catalyzes the removal of the α-amino group from the amino acid threonine as well as the shift of the OH group.

$$\underset{\text{threonine}}{CH_3-\overset{\overset{\displaystyle OH}{|}}{CH}-\overset{\overset{\displaystyle NH_3^+}{|}}{CH}-COO^-} \longrightarrow \underset{\alpha\text{-ketobutyrate}}{CH_3-CH_2-\overset{\overset{\displaystyle O}{||}}{C}-COO^-} + \underset{\text{ammonia}}{NH_3}$$

The amount of α-ketobutyrate product is determined by measuring the NADH consumed in the coupled reaction,

$$\underset{\alpha\text{-ketobutyrate}}{CH_3-CH-\overset{\overset{\displaystyle O}{||}}{C}-COO^-} + NADH + H^+ \longrightarrow \underset{\text{hydroxybutyrate}}{CH_3-CH_2-\overset{\overset{\displaystyle OH}{|}}{CH}-COO^-} + NAD^+$$

The μmol NADH oxidized to NAD$^+$ exactly equals the μmol ketobutyrate produced by threonine dehydratase. Since the [NADH] may easily be quantitated by measuring the decrease in absorbance at 340 nm, the $\dot{v}_0$ values may be determined from the absorbance data given below.

Calculate the K_m and V_{max} for threonine dehydratase from the data given below.

[threonine] (μM),	[NADH] μM	Absorbance at 340 nm	
		0 min	1.0 min
100	100	0.622	0.599
50	100	0.622	0.601
33.3	100	0.622	0.604
20	100	0.622	0.607

Total volume = 1.0 mL

absorbance $= a_m bc$;   for 1 mM NADH, $a_m = 6.22$

To calculate c (concentration) from absorbance data, see Chapter 1, p 31.

Solution:

The amount of NADH that has been oxidized during the reaction (NADH) must be determined for each substrate concentration. For example, for the [S] = 100 μM, A decreases from 0.622 to 0.599 in 1.0 min. Using Beer's law, $A = a_m \times b \times c$, the NADH concentrations before (c_0) and after (c_1) the 1.0-min reaction may be determined, assuming that $a_m = 6.22$ for 1.0 mM solution.

$$A_0 = 0.622 \qquad c_0 = \frac{0.622}{6.22} = 0.100 \text{ m}M \quad \text{or} \quad 100 \text{ }\mu M$$

$$A_1 = 0.599 \qquad c_1 = \frac{0.599}{6.22} = 0.0963 \text{ m}M$$

Then the change in [NADH],

$$\Delta[\text{NADH}] = c_0 - c_1 = 0.100 - 0.0963$$

$$= 0.0037 \text{ m}M \quad \text{or} \quad 0.0037 \text{ }\mu\text{mol/mL}$$

Since the volume is 1.0 mL, then 3.7 nmol of product was formed in 1.0 min.

A faster method for calculating the decrease in [NADH] is by using the equation

$$A = a_m \times b \times c$$

where ΔA is the decrease in absorbance and Δc is the change in [NADH]. For example, in the reaction with 100 μM NADH: ($a_m = 6.22$ for 1.0 mM solution of NADH; $b = 1$ cm.)

$$\Delta c = \frac{\Delta A}{a_m \times b}$$

$$= \frac{0.023}{6.22} = 0.0037 \text{ m}M$$

$$\Delta[\text{NADH}] = 0.00370 \text{ m}M \text{ (}\mu\text{mol/mL)}$$

The volume of the reaction mixture was 1.0 mL and therefore the decrease in NADH in 1.0 min is 0.0037 μmol or 3.7 nmol.

| [S], [threonine] (μM) | Absorbance at 340 nm | | | | |
	0 min	1.0 min	ΔA	ΔNADH (μmol)	v_0 (nmol/min)
100	0.622	0.599	0.023	−0.00370	3.70
50	0.622	0.601	0.021	−0.00337	3.37
33.3	0.622	0.604	0.018	−0.00290	2.90
20	0.622	0.607	0.015	−0.00241	2.41

Since the amount of NADH oxidized per minute exactly equals the amount of α-keto-butyrate (product) formed per minute by threonine dehydratase, ΔNADH/min may be treated as v_0. To determine the K_m and V_{max} from these data, the reciprocal values, 1/[S] and 1/v_0, may be plotted in the Lineweaver–Burk plot.

[S], [threonine] (μM)	v_0 (nmol/min)	1/[S]	$1/v_0$
100	3.70	0.01	0.27
50	3.37	0.02	0.30
33.3	2.90	0.03	0.34
20	2.41	0.05	0.42

When the reciprocal values are plotted, a straight line may be drawn through the points, intersecting the x-axis at approximately -0.045 and the y-axis at approximately 0.21.

$$-\frac{1}{K_m} = -0.045$$

$$K_m = 22 \ \mu M$$

$$\frac{1}{V_{max}} = 0.21$$

$$V_{max} = 4.76 \ \text{nmol/min}$$

Problem 3–10

The enzyme β-galactosidase cleaves the synthetic substrate o-nitrophenyl β-galactoside (o-NPG) to produce o-nitrophenol (o-NP) and galactose. This substrate is used for kinetic studies because the product, o-NP, absorbs light strongly at 420 nm but the substrate, o-NPG, does not absorb at this wavelength. Therefore, the increase in absorbance per minute is directly proportional to the product formed per minute, or v_0. Using this substrate allows the β-galactosidase reaction to be followed easily with a spectrophotometer.

$$o\text{-NPG} + H_2O \rightleftharpoons o\text{-NP} + \text{galactose}$$

β-Galactosidase purified from a fungus was used in kinetic studies and gave the following results. The time course assays were linear for more than 5.0 min.

[o-NPG] (mM)	Absorbance (420 nm) after 1.0 min
1.0	0.09
1.33	0.12
2.0	0.17
2.5	0.20
3.7	0.26

Assuming that the a_m for the product, o-NP, is approximately 0.1 for a 1.0 mM solution at pH 7.3, calculate the v_0 for each concentration of o-NPG, and then determine the K_m for o-NPG and the V_{max} from these data.

Solution

Assume that the absorbance (A) of each solution at 0 min is 0.0 and the A after 1.0 min is due to o-NP formation. The v_0 (μmol o-NP formed/min) values are calculated using Beer's law, $A = a_m \times c \times b$, where A is the absorbance, c the concentration of o-NP in μmol/mL, a_m (extinction coefficient) = 0.1, and b = 1.

The total volume of the assay mixture is 1.0 mL and therefore the μmol o-NP present after 1.0 min exactly equals the initial velocity, v_0, μmol o-NP formed/min.

Absorbance after 1.0 min	[o-NP] (μmol/mL/1.0 min)	v_0 (μmol/min)
0.09	0.9	0.9
0.12	1.2	1.2
0.17	1.7	1.7
0.20	2.0	2.0
0.26	2.6	2.6

Sample calculation: A = 0.09, 0.09 = 0.1 $\times$ c $\times$ 1; c = 0.09/0.10 = 0.9 mM or 0.9 μmol/mL.

Assuming that this enzyme obeys Michaelis–Menten kinetics, the reciprocals of the [S] and v_0 values should yield a straight line when plotted in the Lineweaver–Burk plot.

[S] (mM)	v_0 (μmol/min)	1/[S] (1/mM)	1/v_0 (1/μmol/min)
1.0	0.9	1.0	1.11
1.33	1.2	0.75	0.83
2.0	1.7	0.50	0.59
2.5	2.0	0.40	0.50
3.7	2.6	0.27	0.38

From the Lineweaver–Burk plot, the x intercept is approximately −0.1 and the y intercept is approximately 0.1. Then

$$-\frac{1}{K_m} = -0.1$$

$$K_m = 1.0 \text{ m}M$$

and

$$\frac{1}{V_{max}} = 0.1$$

$$V_{max} = 10 \ \mu\text{mol/min}$$

Inhibition of Enzyme-Catalyzed Reactions

Enzymes are inhibited by many chemical compounds which bind to the enzyme and cause a decrease in its ability to catalyze reactions. If an inhibitor binds to the active site of an enzyme, it prevents the substrate molecule from binding. The inhibition is called *competitive* because the inhibitor molecule competes with the substrate for binding to the enzyme. If a compound binds to a site other than the active site, the inhibition is termed *noncompetitive*. The term *allosteric site* is used to designate certain binding sites on the enzyme which are different than the active site. An *allosteric effector* is a molecule that binds to the allosteric site. Allosteric inhibition is a form of noncompetitive inhibition and is probably the most common in cells.

Note: Small molecules that bind to enzymes are called *ligands*. Substrates, allosteric effectors, and other small molecules that may bind and influence protein structure and function are all *ligands*.

In some cases, an inhibitor binds covalently to the enzyme, at the active site or an allosteric site, and the binding is essentially *irreversible. Only reversible forms of inhibition where the inhibitor molecule is bound by electrostatic attractions, hydrophobic interactions, or other noncovalent bonds* are considered here.

Competitive Inhibition

Competitive inhibitors usually resemble the substrate in some way and many are structural analogs. These types of inhibitors bind to the active site of the enzyme, usually by electrostatic attractions or hydrophobic interactions, so that the inhibitor molecule binding is easily *reversible*.

$$E + I \underset{k_1}{\overset{k_2}{\rightleftharpoons}} EI$$

When a competitive inhibitor (I) binds to an enzyme's active site, EI is formed instead of ES, and in most cases, no product, P, can be formed. Thus in a population of E molecules where a competive inhibitor, I, is present, the rate of product formation will be decreased because some E molecules (but not all) bind I to form EI and are unable to bind S.

The effectiveness of an inhibitor is given by its *inhibition constant, K_i,* which is a measure of the enzyme's affinity for a specific inhibitor molecule.

Since

$$K_i = \frac{[E][I]}{[EI]} \tag{14}$$

a very low K_i indicates that most of the inhibitor molecules are bound to the enzyme, whereas a high K_i indicates poor binding of the inhibitor to the enzyme, so that very little EI is present.

K_i is calculated from plots of data obtained in time course assays where an inhibitor [I] was added to certain assay tubes. Table 3–6 shows data from time course assays performed in the presence and absence of a competitive inhibitor. Figure 3–8b illustrates the effect of the competitive inhibitor in Table 3–6b. Note the decrease in the amount of product formed per time unit in the presence of the inhibitor at lower [S]. For example, at [S] = 25 μM (assay 7), 0.16 μmol P was formed in 60 sec in the absence of inhibitor, whereas only 0.009 μmol P was formed in 60 sec with inhibitor present (assay 13). At higher [S], the effects of the inhibitor were decreased because more S competes with constant [I].

Table 3-6. Time Course Assay Data

a. In the absence of a competitive inhibitor

Assay No.	E (mg)	[S] (μM)	\u03bcmol Product Formed 30 sec	60 sec	90 sec	120 sec
1	0.01	10.0	0.005	0.010	0.015	0.018
2	0.01	11.1	0.0055	0.011	0.016	0.019
3	0.01	12.2	0.006	0.011	0.017	0.021
4	0.01	14.2	0.006	0.012	0.019	0.023
5	0.01	16.6	0.008	0.017	0.025	0.030
6	0.01	20.0	0.007	0.015	0.022	0.026
7	0.01	25.0	0.008	0.016	0.024	0.028
8	0.01	33.3	0.011	0.022	0.033	0.040
9	0.01	50.0	0.014	0.028	—	—
10	0.01	100.0	0.0165	0.033	—	—

b. In the presence of a competitive inhibitor[a]

Assay No.	E (mg)	[I] (μM)	[S] (μM)	\u03bcmol Product Formed 30 sec	60 sec	90 sec
11	0.01	10	16.6	0.003	0.007	0.010
12	0.01	10	20.0	0.004	0.008	0.012
13	0.01	10	25.0	0.004	0.009	0.013
14	0.01	10	33.3	0.006	0.012	0.017
15	0.01	10	50.0	0.008	0.017	0.024
16	0.01	10	100.0	0.011	0.022	0.032

[a] No inhibitor was present in assays 1 through 10.

From the time course data in Fig. 3–8, we can calculate the initial velocities (slope of linear portion of time course curve) at the different [S] in the presence and absence of a constant [I]. These v_0 values and the reciprocals required for the Lineweaver–Burk plot (Fig. 3–9) are shown in Table 3–7.

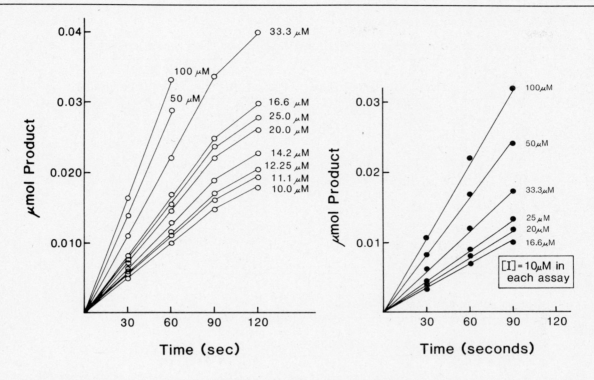

Figure 3-8. Time course assays in the presence and absence of a competitive inhibitor, I (data from Table 3-6). a. no inhibitor present; b. [I] = 10 μM in each assay.

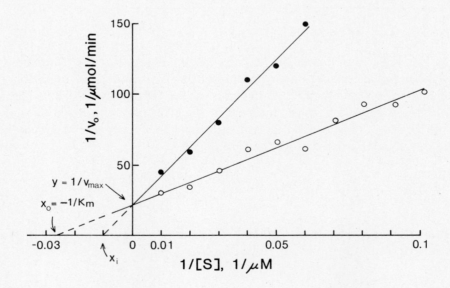

Figure 3-9. Lineweaver-Burk plot. Reciprocal values from time curse data obtained in the presence (10 μM) and absence of a competitive inhibitor (data from Table 3-7). *x_i is the x intercept in the presence of the inhibitor.

Table 3-7. v_0 Values and Reciprocals

a. Initial velocities at different substrate concentrations and their reciprocal values

Assay No.	[I]	[S] (μM)	v_0 (μmol/min)	1/[S] (μM^{-1})	$1/v_0$ (min/μmol)
1	0	10	0.010	0.10	100
2	0	11.1	0.011	0.09	90.9
3	0	12.2	0.011	0.08	90.9
4	0	14.2	0.012	0.07	83.3
5	0	16.6	0.017	0.063	58.8
6	0	20.0	0.015	0.05	66.6
7	0	25.0	0.016	0.04	62.5
8	0	33.3	0.022	0.03	45.4
9	0	50.0	0.028	0.02	35.7
10	0	100.0	0.033	0.01	30.3

b. Inhibited v_0 values and their reciprocals

Assay No.	[I] (μM)	[S] (μM)	v_0 (nmol/min)	1/[S]	$1/v_0$
11	10	16.6	7	0.06	0.141
12	10	20.0	8	0.05	0.125
13	10	25.0	9	0.04	0.111
14	10	33.3	12	0.03	0.083
15	10	50.0	17	0.02	0.059
16	10	100.0	22	0.01	0.045

Note that the V_{max} *is the same in the presence and absence of the competitive inhibitor.* The true K_m remains the same, but in the presence of a competitive inhibitor, the x intercept is decreased (to x_i) by a factor that depends on the [I] and the K_i. K_i *is the dissociation constant for the EI complex.* K_i is a measure of the affinity of the enzyme for the inhibitor molecule, much as K_m may be used as a measure of the affinity of the enzyme for the substrate molecule. K_i is calculated from the formula

$$x_i = -\frac{1}{K_m (1 + [I]/K_i)} \tag{15}$$

Note: A small K_m or K_i (1 μM or less) indicates that the enzyme has very high affinity for the substrate or inhibitor molecule, respectively.

The K_m, V_{max}, and K_i values are determined from the data plotted in Fig. 3–6, as shown below. The x intercept, x_0 (no inhibitor present) $= -1/K_m$. Then $x_0 = -0.26 = -1/K_m$; $K_m = 38.5$ μM.

The y intercept is approximately 24. Then $y = 24 = 1/V_{max}$; $V_{max} = 0.04$ μmol/min. The amount of enzyme used in each assay was 0.01 mg, and if V_{max} is expressed per milligram of enzyme,

$$V_{max} = 0.04 \ \mu\text{mol/min}/0.01 \ \text{mg} = 0.04 \times 100$$
$$= 4 \ \mu\text{mol/min/mg enzyme}$$

The inhibited x intercept,

$$x_i = -\frac{1}{K_m(1 + [I]/K_i)}$$
$$= -0.01$$

[I] = 10 μM and from the calculations above, $K_m = 38.5$ μM. Then

$$-0.01 = -\frac{1}{38.5(1 + 10/K_i)}; \quad K_i = 6.26 \ \mu M$$

(See Problem 1-19, p 17 to solve this equation for K_i)

Noncompetitive Inhibition

Some inhibitor molecules bind to enzymes *not at the active site,* but rather, *at allosteric sites.* Binding at an allosteric site causes conformational changes in the protein which may alter the active site and decrease the rate of product formation. This type of inhibition of enzyme activity is termed *noncompetitive* inhibition because the inhibitor molecules are not competing with substrate molecules for binding. Noncompetitive inhibition is found at the beginning of almost all metabolic pathways in cells. It is the most common mechanism for controlling the rate of enzyme activity in cells. Generally, the noncompetitive inhibitor molecule (*allosteric effector*) is quite different, structurally, from the substrate.

Noncompetitive (or allosteric) inhibitors decrease the activity of the enzyme when bound to the allosteric site of the enzyme. Increasing the substrate concentration will not relieve the noncompetitive type of inhibition because the substrate and the inhibitor bind to different sites. Therefore, V_{max} cannot be attained simply by increasing [S] as seen in competitive inhibition.

Classic Form of Noncompetitive Inhibition. In the *classic* form of *noncompetitive inhibition,* the *binding of the substrate to the active site is not affected* by the inhibitor binding to the allosteric site, even though the reaction rate decreases. Data from time course assay done in the presence and absence of a class noncompetitive inhibitor are plotted in Fig. 3-10.

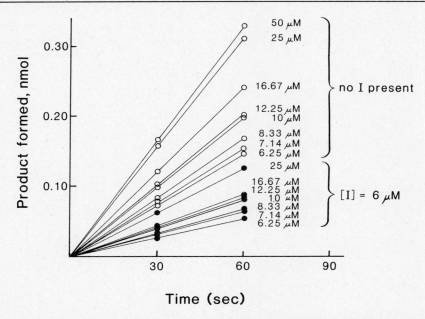

Figure 3–10. Time course assay data in the presence and absence of a noncompetitive (classic) inhibitor.

Initial velocities calculated from the time course data and reciprocals are presented in Table 3–8 and are plotted in Fig. 3–11. When $1/v_0$ is plotted versus $1/[S]$, the V_{max} is $1/v_0$ and the K_m is $-1/x$, as shown in Fig. 3–11 for a classic noncompetitive inhibitor.

Table 3–8. **Data Obtained from Time Course Assays in the Presence and Absence of a Classic Noncompetitive Inhibitor**

a. No inhibitor present

[S] (μM)	v_0 (μmol/min)	1/[S] (1/μM)	$1/v_0$ (1/μmol/min)
50	0.33	0.02	3.0
25	0.31	0.04	3.2
16.6	0.24	0.06	4.1
12.2	0.19	0.08	5.1
10	0.20	0.10	4.9
8.33	0.16	0.12	6.0
7.14	0.14	0.14	6.8
6.25	0.153	0.16	6.5

b. Classic noncompetitive inhibitor present, [I] = 6 μM in each assay

[S]	v_0 (μmol/min)	1/[S] (1/μM)	1/v_0 (1/μmol/min)
50	0.153	0.01	6.5
25	0.125	0.04	8.0
16.6	0.09	0.06	11.0
12.2	0.086	0.08	11.6
10	0.081	0.10	12.4
8.33	0.067	0.12	15.0
7.14	0.063	0.14	15.9
6.25	0.055	0.16	18.0

When the data in Table 3–8b are plotted in Fig. 3–11, the line drawn through the points intersects the *x*-axis at the same point as the line through the uninhibited $1/v_0$ values. In classic noncompetitive inhibition the inhibitor binding to the enzyme does not change the K_m, the enzyme's affinity for the substrate.

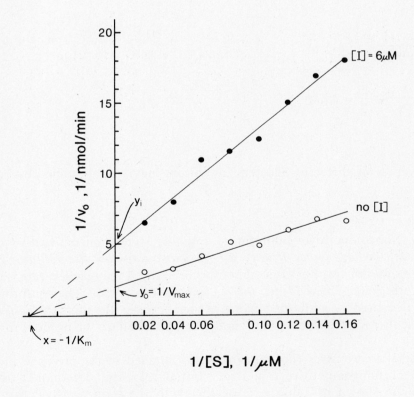

Figure 3–11. Lineweaver–Burk plot of reciprocal values in Table 3–8, showing classic form of noncompetitive inhibition.

From the double-reciprocal plot in Fig. 3–11, we see that the y-intercept is increased, and therefore V_{max} *is decreased* in the presence of the inhibitor. With no inhibitor:

$$V_{max} = \frac{1}{y_0} = \frac{1}{2} = 0.5 \text{ nmol/min/mg}$$

In the presence of the inhibitor:

$$\text{apparent } V_{max} = \frac{1}{y_i} = \frac{1}{5} = 0.2 \text{ nmol/min/mg}$$

K_m *is not changed in the presence of a classic noncompetitive inhibitor.*

$$K_m = -\frac{1}{x_0} = \frac{1}{0.06} = 16.6 \text{ } \mu M$$

K_i represents the affinity of the enzyme's *allosteric site* for binding the noncompetitive inhibitor. K_i is calculated as

$$K_i = \frac{y_0 [I]}{y_i - y_0} \tag{16}$$

$$= \frac{2 \times 6}{5 - 2} = \frac{12}{3} = 4 \text{ } \mu M$$

The effectiveness of different allosteric inhibitors may be determined by comparing their K_i values.

Mixed-Type Noncompetitive Inhibition.   In mixed-type noncompetitive inhibition, the *enzyme appears to have lower affinity for the substrate when the inhibitor is bound to the allosteric site.* In this form of noncompetitive inhibition, the reaction rate is decreased as in the classic form, but *in mixed type, the K_m apparent increases.* The conformational change in the protein caused by inhibitor binding to the allosteric site alters the active site so that it is more difficult for substrate to bind, and when it is bound, the product forms at a slower rate.

Time course assay curves (not shown) were obtained in the presence and absence of a mixed type of noncompetitive inhibitor. Initial velocities determined from these time course plots are shown in Table 3–9 together with the reciprocals values for v_0 and [S]. When $1/v_0$ is plotted versus $1/[S]$, the V_{max} is $1/y_0$ and the K_m is $-1/x$, as shown in Fig. 3–12.

Table 3-9. Data Obtained from Time Course Assays in the Presence and Absence of a Noncompetitive Inhibitor, I, Giving Mixed-Type Inhibition

a. No inhibitor present

[S] (μM)	v_0 (nmol/min)	1/[S] (μM^{-1})	1/v_0 (nmol^{-1} min)
6.25	0.15	0.16	6.8
7.14	0.15	0.14	6.8
8.33	0.17	0.12	6.0
10.00	0.20	0.10	5.0
12.25	0.19	0.08	5.2
16.60	0.24	0.06	4.1
25.00	0.32	0.04	3.1
50.00	0.33	0.02	3.0

b. Mixed-type noncompetitive inhibitor present at 6 μM in each assay

[S] (μM)	v_0 (nmol/min)	1/[S] (μM^{-1})	1/v_0 (nmol^{-1} min)
6.25	0.047	0.16	21.2
7.14	0.015	0.14	19.6
8.33	0.055	0.12	18.0
10.00	0.067	0.10	15.0
12.25	0.081	0.08	12.4
16.60	0.083	0.06	12.0
25.00	0.11	0.04	9.2
50.00	0.13	0.02	7.7

When the data in Table 3-9b are plotted in Fig. 3-12, the line drawn through the points intersects the x-axis (x_I) at a point different from the x intercept of the line representing uninhibited v_0. This shows that the inhibitor binding to the enzyme does change the K_m.

Figure 3-12 shows that the mixed type of noncompetitive inhibitor decreases the apparent V_{max} and increases the apparent K_m, indicating lowered reaction rates and substrate binding affinities, respectively, in the presence of the inhibitor. With no inhibitor:

$$true\ V_{max} = \frac{1}{y_0} = \frac{1}{2} = 0.5\ \mu mol/min/mg\ enzyme$$

In the presence of the mixed-type noncompetitive inhibitor:

$$apparent\ V_{max} = \frac{1}{y_i} = \frac{1}{5} = 0.2\ \mu mol/min/mg\ enzyme$$

With no inhibitor:

$$true \ K_m = -\frac{1}{x_0} = \frac{1}{0.06} = 16.6 \ \mu M$$

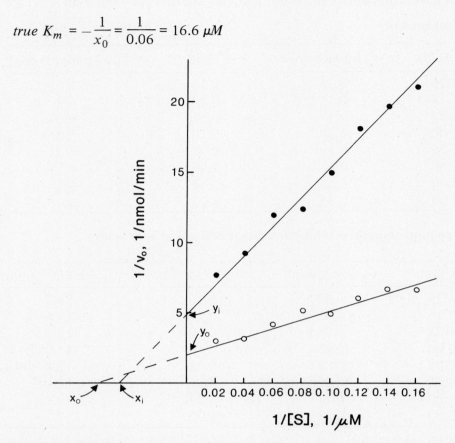

Figure 3-12. Lineweaver–Burk plot of data obtained in the presence (●–●) and absence (○–○) of a mixed-type noncompetitive inhibitor.

In the presence of a mixed-type noncompetitive inhibitor, the apparent K_m is increased:

$$apparent \ K_m = \frac{-1}{0.05} = 20 \ \mu M$$

The K_i is a measure of the affinity of the enzyme for the inhibitor. In mixed-type non-competitive inhibition, the slope of *the line through inhibited points* is

$$slope = \frac{K_m}{V_{max}} \left(1 + \frac{[I]}{K_i}\right) \tag{17}$$

The slope is determined by calculating the run/rise, which from Fig. 3–12 is 15/0.15. Therefore, the slope of the inhibited line = 100. Since (true) K_m = 16.6 μM, (true) V_{max} = 0.5 μmol/min/mg, and [I] = 10 μM, the expression above becomes

$$100 = \frac{16.6}{0.5}\left(1 + \frac{10\ \mu M}{K_i}\right)$$

$$100 = 33.2\left(1 + \frac{10\ \mu M}{K_i}\right)$$

$$3.0 = 1 + \frac{10\ \mu M}{K_i}$$

$$2.0 = \frac{10}{K_i}$$

$$K_i = \frac{10}{2.0} = 5.0\ \mu M$$

Problem 3-11

In Table 3-10, time course assay data are given for an enzyme with no inhibitor present (tubes 1-6). The same enzyme assays were repeated with inhibitor A (I_A) present at 20 μM (tubes 7-10). From the time course assay data below, determine which type of inhibition is found with inhibitor A and calculate the K_m and K_i values.

Table 3-10. Time Course Assay Data in the Absence and Presence of Inhibitor A (I-A)

Tube No.	[I] (μM)	[E] (mg)	[S] (μM)	nmol P Formed per Time Unit			
				30 sec	60 sec	90 sec	120 sec
1	—	0.01	1.67	0.65	1.30	1.51	1.60
2	—	0.01	2.00	0.81	1.62	1.80	1.90
3	—	0.01	3.33	0.98	1.96	2.30	2.55
4	—	0.01	5.00	1.16	2.23	3.00	—
5	—	0.01	10.00	1.24	2.5	—	—
6	—	0.01	20.00	1.62	3.24	—	—
7	20	0.01	3.33	0.42	0.84	—	1.13
8	20	0.01	5.00	0.54	1.09	—	1.49
9	20	0.01	10.00	0.83	1.67	—	1.98
10	20	0.01	20.00	1.12	2.22	—	3.15

Solution:

This problem is solved exactly as in Problem 3–5. First *plot the time course data* (given in Table 3–10). Figure 3–13a shows the time course data obtained in the absence of any inhibitor. Figure 3–13b is time course data obtained in the *presence* of inhibitor A with [I-A] = 20 μM.

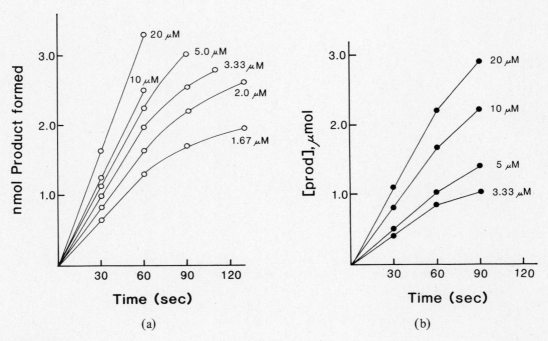

Figure 3–13.

Now *calculate an initial velocity for each [S] in the presence and absence of inhibitor A*. Find the reciprocal values for each [S] and v_0 (Table 3–11).

Table 3–11. Initial Velocities in the Presence and Absence of Inhibitor A

a. [I] = 0

[S] (μM)	v_0 (nmol/min)	1/[S] (μM^{-1})	$1/v_0$
1.67	1.30	0.59	0.77
2.00	1.62	0.50	0.62
3.33	1.96	0.30	0.51
5.00	2.32	0.20	0.43
10.00	2.50	0.10	0.40
20.00	3.24	0.05	0.31

b. [I] = 20 μM

[S] (μM)	v_0 (nmol/min)	1/[S] (μM^{-1})	$1/v_0$
3.33	0.84	0.30	1.19
5.00	1.09	0.20	0.91
10.00	1.67	0.10	0.60
20.00	2.22	0.05	0.45

Plot the reciprocals $1/v_0$ versus 1/[S] (from Table 3–11) as shown in Fig. 3–14. Draw a straight line through both sets of points.

In this case the lines intersect on the y axis, indicating that V_{max} is not changed in the presence of the inhibitor. Therefore, the *type of inhibition caused by inhibitor A is competitive*. The true K_m is calculated as shown below from x_0:

$$K_m = -\frac{1}{x_0} = -\frac{1}{-0.4} = 2.5 \ \mu M$$

The K_i may be calculated as shown below from the inhibited x intercept, x_i:

$$x_i = -\frac{1}{K_m \ (1 + [I]/K_i)} ; \quad -0.1 = -\frac{1}{2.5(1 + 20/K_i)} ; \ K_i = 6.67 \ \mu M$$

where x_i is the x intercept in the presence of inhibitor A and x_0 is the x intercept in the absence of inhibitor.

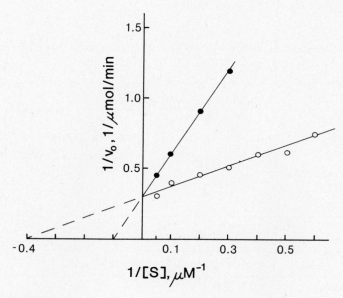

Figure 3–14. Lineweaver–Burk plot.

Problem 3-12

In Table 3-12 time course data are given for the enzyme with no inhibitor (tubes 1-6) and with inhibitor B (I-B) at 15 μM (tubes 7-10). From these data, determine which type of inhibition is caused by I-B and calculate the true K_m and the K_i.

Table 3-12. Time Course Assay Data in the Absence and Presence of Inhibitor B

Tube No.	$[I_B]$ (μM)	[E] (mg)	[S] (μM)	nmol P Formed per Time Unit			
				30 sec	60 sec	90 sec	120 sec
1	—	0.01	1.67	0.65	1.30	1.51	1.60
2	—	0.01	2.00	0.81	1.62	1.80	1.90
3	—	0.01	3.33	0.98	1.96	2.30	2.55
4	—	0.01	5.00	1.16	2.23	3.00	—
5	—	0.01	10.00	1.24	2.5	—	—
6	—	0.01	20.00	1.62	3.24	—	—
7	15	0.01	2.00	0.41	0.83	—	1.24
8	15	0.01	3.30	0.53	1.05	—	1.58
9	15	0.01	5.00	0.62	1.25	—	1.87
10	15	0.01	10.00	0.75	1.51	—	2.26

Solution:

Plot the data for time course assays done in the presence and absence of inhibitor B (Table 3-12). Follow the steps in the solution to Problem 3-11, calculating v_0 values determined from the time course assays plots. Plot the reciprocal values (1/[S] versus $1/v_0$) and draw a straight line through both sets of data points.

Table 3-13. Initial Velocities in the Presence and Absence of Inhibitor B

a. [I] = 0

[S] (μM)	v_0 (nmol/min)	1/[S] (μM^{-1})	$1/v_0$
1.67	1.30	0.59	0.77
2.00	1.62	0.50	0.62
3.33	1.96	0.30	0.51
5.00	2.32	0.20	0.43
10.00	2.50	0.10	0.40
20.00	3.24	0.05	0.31

b. $[I_B]$ = 15 μM

[S] (μM)	v_0 (nmol/min)	1/[S] (μM^{-1})	$1/v_0$
2.00	0.83	0.50	1.20
3.33	1.05	0.30	0.95
5.00	1.25	0.20	0.80
10.00	1.51	0.10	0.66

The lines do not intersect on the y-axis, which shows that the V_{max} is altered in the presence of inhibitor B. The lines do intersect on the x-axis, indicating that the K_m is unchanged in the presence of B. This type of inhibition is the classic form of noncompetitive inhibition (compare to Fig. 3–8b). K_m is calculated from the x intercept.

$$x = -0.4 = -\frac{1}{K_m} \quad K_m = 2.5 \ \mu M$$

K_i may be calculated:

$$K_i = \frac{y_0 [I]}{y_i - y_0} \tag{16}$$

$$= \frac{0.30 \times 15 \ \mu M}{0.55 - 0.30} = \frac{4.5}{0.25} = 18 \ \mu M$$

Summary for All Types of Enzyme Inhibition

1. The *true K_m* is calculated in the absence of inhibitors. An *apparent K_m* is calculated in the presence of inhibitors and is greater than the *true K_m* in *competitive inhibition*. In *noncompetitive inhibition,* the apparent K_m *may be greater than the true K_m* (mixed type) but in many cases has been found to be the same (classic form).

2. V_{max} is calculated in the absence of inhibitors. In the presence of a competitive inhibitor, V_{max} is *always the same* as in its absence. In the presence of a *noncompetitive inhibitor,* an apparent V_{max} is calculated and is always less than the true V_{max}.

Reactions with Two Substrates

Many of the enzyme-catalyzed reactions that occur in cells require two different substrates to form the product. In these cases, each substrate binds to the active site during the reaction. One type of two-substrate reaction may be written

$$E + S_1 \rightleftharpoons E - S_1 \rightleftharpoons \quad E - S_1 + S_2 \rightleftharpoons \underset{\underset{S_2}{|}}{E - S} \longrightarrow E - P$$

where S_1 designates one substrate, S_2 designates the other substrate, and S_2-E-S_1 represents the condition where both substrates are bound to the active site.

Although one of the assumptions made by Michaelis and Menten was that only one substrate molecule is bound by one enzyme molecule, K_m values for each S can be measured in the same way by imposing conditions that mimic Michaelis–Menten conditions.

When the S_1 concentration is so high that the S_1 site is always filled (saturation conditions), the reaction above becomes

$$ENZ\!-\!S_1 + S_2 \rightleftharpoons \underset{\underset{\textstyle S_2}{|}}{ENZ}\!-\!S_1 \longrightarrow ENZ + P$$

and we are measuring the initial velocity as a function of the S_2 concentration only. V_{max} is reached *only* when *both* substrates are at very high concentrations (more than $10 \times K_m$).

Note: In many cases, one of the substrates is not soluble at the high levels required for saturation. In this case, more complex calculations are required using both $[S_1]$ and $[S_2]$ to determine the V_{max} and both K_m values.

Consider the following situation. The enzyme hypoxanthine phosphoribosyl transferase (HPRTase) binds the substrates phosphoribosyl pyrophosphate (PRPP) and hypoxanthine to separate parts of its active site, and catalyzes the formation of the products inosine monophosphate (IMP) and pyrophosphate (PP_i).

hypoxanthine   PRPP   IMP PP_i

In time course assays, [PRPP], S_1, was 5.0 μM, more than 50 times its K_m value and therefore at saturation levels. The amount of the product, IMP, formed at different concentrations of S_2, hypoxanthine, was measured and is plotted as a function of time. From this time course data (not shown), an initial velocity was determined for each [S] by taking the slope of the linear portion of the curve.

The K_m and the V_{max} for hypoxanthine were determined from these data by plotting the reciprocal values for [hypoxanthine] and v_0 as shown in Fig. 3–15. Similarly, initial velocities for this same reaction may be calculated using the Michaelis–Menten equation when values for V_{max} and the K_m for each substrate have been determined.

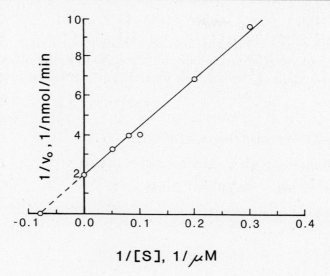

Figure 3-15. Double-reciprocal plots of v_0 values at varying [hypoxanthine]. The [PRPP] was 5.0 mM in each assay. The K_m for hypoxanthine is determined from the x intercept: $x = -0.08 = -1/K_m$; $K_m = 12.5 \mu M$. The V_{max} is calculated from the y intercept: $y = 2 = 1/V_{max}$; $V_{max} = 0.5$ nmol/min/0.1 mg protein = 5 nmol/min/mg.

Problem 3-13

Using the values K_m for hypoxanthine = 12.5 μM and V_{max} = 5.0 nmol/min/mg from the example in Fig. 3-15, determine the amount of IMP formed if 1.0 mg of the enzyme HPRTase reacted under the following conditions:

[hypoxanthine] = 20 μM [PRPP] = 4.0 mM K_m for PRPP = 50 μM

Solution:

The substrate that is saturating is PRPP, since its concentration is more than 50 times its K_m. Therefore, the initial velocity depends on the concentration of the other substrate, [hypoxanthine]. Substituting in the Michaelis–Menten equation, we obtain

$$v_0 = \frac{0.005 \ \mu mol/min/mg \times 2}{12.5 + 20}$$

$$= \frac{0.01}{32.5}$$

$$= 0.0003 \ \mu mol/min/mg$$

Problem 3–14

An enzyme that is found in many human cells is adenosine kinase. This enzyme catalyzes the transfer of a phosphate group from one substrate, ATP, to the other substrate, adenosine, to form ADP and adenylate. Calculate how much adenylate will be formed under the following conditions.

adenosine + ATP $\longrightarrow$ adenylate + ADP

K_m for adenosine = 0.5 μM [adenosine] = 1.0 μM

K_m for ATP = 12 μM [ATP] = 1.0 mM

V_{max} = 0.1 μmol/min/mg protein

Solution:

The substrate that is at saturation levels is the ATP since the [ATP] is greater than 50 times its K_m. Therefore, the v_0 of this reaction depends on the concentration of the other substrate, adenosine.

The Michaelis–Menten equation may be used to determine v_0, where [S] = [adenosine] = 1.0 μM.

K_m for adenosine = 0.5 μM

V_{max} = 0.1 μmol/min/mg protein

$$v_0 = \frac{V_{max} \times [S]}{K_m + [S]}$$

$$= \frac{0.1 \ \mu\text{mol/min/mg} \times 1.0 \ \mu M}{0.5 \ \mu M + 1.0 \ \mu M}$$

$$= \frac{0.1}{1.5}$$

$$= 0.67 \ \mu\text{mol/min/mg enzyme}$$

Problem 3–15

Glucose-6-phosphate dehydrogenase is an enzyme that catalyzes the oxidation of G-6-P to phosphogluconate.

G-6-P + NADP$^+$ $\longrightarrow$ 6-phosphogluconate + NADPH + H$^+$

The cofactor $NADP^+$ accepts a hydride ion from G-6-P in the reaction to form NADPH. Since NADPH absorbs light strongly at a wavelength of 340 nm and $NADP^+$ does not, the amount of NADPH product formed was followed spectrophotometrically. G-6-P dehydrogenase (G-6-PD) purified from a bacterial extract gave the results shown in Table 3–14 in a kinetic study with saturating concentrations of $NADP^+$. Determine the K_m for G-6-P and the V_{max} from these data. ATP was tested as an inhibitor of 6-phosphogluconate formation, and the v_0 values measured in the presence of 1.0 mM ATP are presented in Table 3–7. What type of inhibition did 1.0 mM ATP cause, and what is the K_i for ATP?

Table 3–14. Initial Velocities of G-6-PDH Activity in the Presence and Absence of the Inhibitor ATP

[G-6-P] (mM)	v_0 (μmol NADPH formed/min/μg enzyme[a])	
	No ATP Present	1.0 mM ATP Present
0.040	11	5
0.050	13	6
0.067	16	13
0.100	20	16
0.200	28	20

[a]The μmol NADPH formed should exactly equal the μmol phosphogluconate formed in the reaction.

Solution:

Reciprocals of [S] and v_0 are calculated and shown in Table 3–15. These reciprocal values were plotted in the Lineweaver–Burk plot and straight lines were drawn through the points. The uninhibited x intercept was approximately -7.5. Thus

$$-7.5 = -\frac{1}{K_m}$$

$$K_m = 133 \ \mu M$$

Table 3–15. Reciprocal Values for v_0 and [S]

[G-6-P] (mM)	1/[S]	No ATP Present		1.0 mM ATP Present	
		v_0 (μmol/min/μg)	$1/v_0$	v_0 (μmol/min/mg)	$1/v_0$
0.040	25	11	0.090	5	0.20
0.050	20	13	0.075	6	0.16
0.067	15	16	0.062	13	0.13
0.100	10	20	0.050	16	0.09
0.200	5	28	0.035	20	0.06

The y intercept was approximately 0.02. Thus

$$\frac{1}{V_{max}} = 0.02$$

$$V_{max} = 50 \ \mu mol/min/\mu g \ enzyme$$

The lines intersected on the y-axis indicating that *ATP was a competitive inhibitor of G-6-P* for binding to the active site. The x intercept in the presence of 1.0 mM ATP was about −3. Since $x_i = -1/K_m (1 + [I]/K_i)$, thus

$$-3 = -\frac{1}{0.133(1 + 1.0 \ mM/K_i)}$$

$$3(0.133 + 0.133/K_i) = 1$$

$$0.399 + \frac{0.399}{K_i} = 1$$

$$K_i = 0.66 \ mM$$

Problem 3–16

The enzyme lactate dehydrogenase catalyzes the reaction

$$lactate + NAD^+ \longrightarrow pyruvate + NADH + H^+$$

The cofactor, NAD^+, does not absorb light at 340 nm, but the reduced form, NADH, absorbs strongly at this wavelength. The absorbance of the product NADH is directly related to its concentration, as shown in Chapter 1. Therefore, we can determine the amount of NADH produced and thus initial velocities in enzyme-catalyzed reactions by reading the absorbance of the assay mixture exactly 1.0 min after the reaction has begun. (This method is valid if the time course curves are linear and the [lactate], the second substrate, is at saturating levels.)

For example, if the dehydrogenase enzyme is mixed with the substrates NAD^+ and lactate at time 0 min, the absorbance of the solution (at 340 nm) should read 0.00 at 0 min. At exactly 1.0 min, the absorbance at 340 nm must be due to the product, NADH, formed in the reaction. By using the equation $A = a_m \times b \times c$ relating the [NADH] to its absorbance, we can determine how much product is formed in 1.0 min (i.e., the initial velocity, v_0).

From the absorbances listed in Table 3–16, read at 1.0 min:

(a) Determine the v_0 values for each [S].

(b) Plot the reciprocal values ($1/v_0$ versus $1/[S]$)

(c) From the $1/v_0$ versus $1/[S]$ plot, calculate the V_{max} and the K_m for NADH.

Table 3–16. Absorbances of Assay Mixtures after 1.0 min at 340 nm

	Absorbance at:	
[S] (μM)	0 min	1.0 min
200	0	0.194
100	0	0.154
66	0	0.118
50	0	0.098
40	0	0.083

Note: The time course curves were linear for at least 2.0 min and the [lactate] was 30 times its K_m (saturation levels). Assume that the total reaction mixture has a volume = 1.0 mL and that 0.02 mg of enzyme was used in each assay.

Solution:

(a) Since the time course curves were linear for more than 1.0 min, we can convert the absorbance values at 1.0 min to μmol NADH formed/min and use these values as initial velocity values. The v_0 for each [S] are shown in Table 3–17. These were determined from the formula $A = a_m \times b \times c$, where a_m = 6220 or 6.220 for 1 mM solution of NADH (Chapter 1). For example,

$$A = 0.194 = 6.22 \times 1c; c = 0.031 \text{ m}M \text{ or } 0.031 \ \mu\text{mol/mL}$$

Since the assay volume was 1.0 mL, 0.031 μmol was formed/per minute.

Table 3–17. v_0 Values Determined from Absorbance after 1.0 min, Using the Extinction Coefficient, for NADH

[S] (μM)	v_0 (μmol/min)	$1/[S]$ (μM^{-1})	$1/v_0$ (μmol/min^{-1})
200	0.031	0.005	32
100	0.024	0.010	42
66	0.019	0.015	53
50	0.015	0.020	67
40	0.013	0.025	77

(b) Plots of $1/v_0$ versus $1/[S]$ gave an approximate x intercept at -0.008 and an approximate y intercept at 18.

(c) $$y = \frac{1}{V_{max}} = 18$$

$V_{max} = 0.055\ \mu mol/min/0.02$ mg enzyme

$V_{max} = 2.75\ \mu mol/min/mg$ enzyme

$$x = -\frac{1}{K_m} = -0.008$$

$K_m = 125\ \mu M$

Problem 3-17

The dehydrogenase enzymes that use the cofactor NAD^+ to accept a H^- ion (hydride ion) from the substrate also donate a proton (H^+) to the assay mixture. For example:

$$CH_3-CH_2OH + NAD^+ \longrightarrow CH_3-CHO + NADH + H^+$$

The enzyme alcohol dehydrogenase binds ethanol and catalyzes the removal of a hydride ion and a proton to give the product acetaldehyde. The hydride ion is bound to the NAD^+ to give NADH and *the proton is released to the assay mixture.*

Instead of measuring the absorbance of the product NADH, assume that the amount of H^+ ion formed was measured as a pH change after 1.0 min to determine the initial velocity (i.e., the amount of product formed per minute).

(a) From the data given in Table 3-18, calculate the nanomoles of product (H^+ ion) formed/min (v_0) for each ethanol concentration [S].

(b) Using these v_0 values, determine K_m and V_{max} for this dehydrogenase.

Table 3-18. pH Changes after 1.0 min at Various [S] [a]

Assay	[S] (μM)	pH at: 0 min	pH at: 1.0 min
1	100	7.00	6.89
2	50	7.00	6.91
3	25	7.00	6.94
4	20	7.00	6.95

[a] Total volume of each assay mixture = 1.0 mL. Each assay tube contained 0.01 mg of enzyme preparation.

Solution:

(a) The amount of product formed in 1.0 min may be determined by subtracting the amount of H^+ ions at 0 min from the amount formed after 1.0 min of reaction. We can calculate each $[H^+]$ from the pH values since *the pH is the negative log of the $[H^+]$*.

[S]	Amount of $[H^+]$ formed
100 nmol/mL	$10^{-6.89} - 10^{-7.00}$

$$= \text{antilog} (-6.89) - \text{antilog} (-7.00)$$

$$= (1.29 \times 10^{-7}) - (1.0 \times 10^{-7}) = 0.29 \times 10^{-7} \text{ mol/L}$$

$$= 0.029 \text{ nmol/mL}$$

Since the assay volume = 1.0 mL, 0.029 nmol of H^+ was formed per/minute.

50 nmol/mL: $10^{-6.91} - 10^{-7.00}$

$\qquad\qquad = 0.23 \times 10^{-7} \quad = 0.023$ nmol

25 nmol/mL: $10^{-6.94} - 10^{-7.00} = 0.115 \times 10^{-7}$

$\qquad\qquad = 0.012$ nmol

20 nmol/mL: $10^{-6.95} - 10^{-7.00} = 0.112 \times 10^{-7}$

$\qquad\qquad = 0.011$ nmol

(b) Using these v_0, we can plot $1/v_0$ versus $1/[S]$:

[S] (μM)	v_0 (nmol/min)	$1/[S]$ ($1/\mu M$)	$1/v_0$ (1/nmol)
100	0.029	0.01	34.5
50	0.023	0.02	43.5
25	0.012	0.04	83.5
20	0.011	0.05	90.9

From the plot of $1/v_0$ versus $1/[S]$, a straight line may be drawn which intersects the *x*-axis at about -0.01 and the *y*-axis at about 17. Thus

$$-\frac{1}{K_m} = \frac{1}{-0.01}$$

$$K_m = 100 \; \mu M$$

$$\frac{1}{V_{max}} = 17$$

$$V_{max} = 0.058 \text{ nmol/min/0.01 mg enzyme}$$

Problem 3-18

The enzyme alcohol dehydrogenase is present in most cells and catalyzes the oxidation of ethanol to acetaldehyde. NAD^+ is required to accept the hydride ion removed from ethanol.

$$CH_3CH_2OH + NAD^+ \longrightarrow CH_3CHO + NADH + H^+$$

ethanol   acetaldehyde

Recently, Mardh et al. showed that the steroid hormone testosterone binds to this enzyme and inhibits its activity. Data replotted from their paper are shown in Fig. 3-16. From these data, calculate the true V_{max}, the true K_m for both ethanol and NAD^+, and the K_i for testosterone. What type of inhibition does testosterone cause?

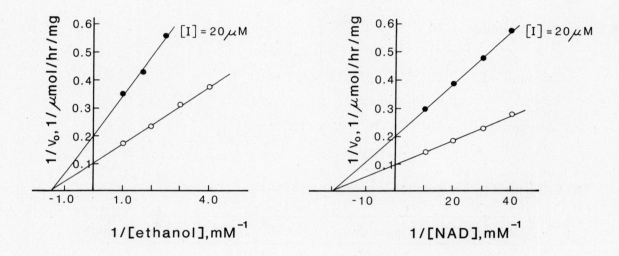

Figure 3-16. Inhibition of alcohol dehydrogenase activity by testosterone. The concentration of NAD^+ was at saturation levels for all data points in A. The concentration of ethanol was at saturation for all data points in B. [From Mardh et al., 1986, *Proc. Natl. Acad. Sci. (USA)* 83:2836–2839.]

Solution:

The true V_{max} is calculated from the y intercept in the absence of any inhibitor. The y intercept is approximately 0.1 from both graphs A and B. If $1/V_{max} = 0.1$, $V_{max} = 10$ μmol/hr/mg. Then the true V_{max} is 10 μmol/hr/mg protein.

The K_m for ethanol is calculated from the x intercept in graph A, which is -1.5. Then $-1/K_m = -1.5$ and $K_m = 1/1.5$.

K_m for ethanol = 0.67 mM

The x intercept from graph B, -25, is the reciprocal of the K_m for NAD^+: $-1/K_m = -25$ and $K_m = 1/25$.

K_m for NAD^+ is 0.04 mM

The type of inhibition caused by testosterone is *noncompetitive since the V_{max} is lowered in the presence of the inhibitor.* The K_m remains the same with inhibitor present, and therefore this is the classic form of noncompetitive inhibition. The formula for determining the K_i for this type of inhibitor is

$$K_i = \frac{y_0 [I]}{y_i - y_0}$$

Using the data from graph B, where $[NAD^+]$ was varied,

$$y_0 = 0.1 \quad \text{when } [I] = 20 \ \mu M, y_i = 0.20$$

Then $K_i = 0.20 \times 20 \ \mu M/(0.20 - 0.1) = 40 \ \mu M$ for testosterone as an allosteric inhibitor of alcohol dehydrogenase.

Enzyme Specificity

By determining the specificity of enzymes, biochemists can sometimes design molecules that inhibit a specific enzyme's activity and prevent or decrease symptoms of disease. The specificity of an enzyme may be defined as those chemical groups on the substrate that are required for binding to the active site. Some enzymes bind many structurally different substrates (e.g., some esterases) and some enzymes bind only one molecule as a substrate (e.g., PEP carboxylase). Most enzymes bind some structural analogs of the substrate as well as the substrate itself.

To determine the specificity of substrate binding, structurally related compounds may be used as potential substrates and K_m values may be calculated. However, many compounds structurally related to the substrate may bind to the active site but cannot be converted to product. In these instances, the structural analogs are used as potential competitive inhibitors of substrate binding. The K_i values are calculated and compared to the K_m value for the substrate. Low K_i values indicate high affinity of the enzyme for the inhibitor, whereas high K_i values indicate low binding affinity.

Consider as an example the enzyme xanthine oxidase, which catalyzes the formation of uric acid from the purine bases hypoxanthine or xanthine in humans.

hypoxanthine   uric acid   xanthine   uric acid

The K_m for hypoxanthine is approximately 15.0 μM, and for xanthine, the K_m was found to be 45 μM.

A few compounds used as competitive inhibitors of the normal substrate hypoxanthine are shown in Table 3-19 with their K_i values. The hypoxanthine analog, adenine, was used as a competitive inhibitor in time course assays and the K_i was greater than 800 μM. Adenine differs from hypoxanthine only in the amino group on carbon 6. The $-NH_2$ group is similar in size to the $-OH$ group but has very different chemical properties which prevent adenine from binding efficiently to the active site of xanthine oxidase.

Table 3-19. K_i Values for Potential Competitive Inhibitors of Xanthine Oxidase[a]

Potential Competitive Inhibitor		K_i value
Purine		900
Adenine		800
Guanine		300
Allopurinol		38
Inosine		900

[a] The K_m for hypoxanthine is 15 μM and for xanthine is 45 μM.

Comparing the structures of competitive inhibitors and their K_i values to the substrates hypoxanthine and xanthine and their K_m values, it is clear that the enzyme requires purine bases with —OH groups at C_2 or C_6. The enzyme binds substrates with an —OH group at C_2, but not an amino group attached to C_2 as in guanine. The addition of a sugar to the N_9 position prevents binding, as shown by the high K_i value for inosine. Changing N_7 to a carbon and C_8 to a nitrogen atom does not prevent allopurinol from binding to the active site, as shown by the low K_i value for allopurinol.

The hypoxanthine analog allopurinol is an effective competitive inhibitor of hypoxanthine, but allopurinol is not converted to any product. Since allopurinol is not toxic, it has been used to treat millions of people who produce too much uric acid. These individuals, when untreated, suffer from an extremely painful form of arthritis known as gout, in which the tiny needle-like crystals of uric acid precipitate out of the bloodstream and lodge in the joints, especially of the feet. Allopurinol binds to the active site of the xanthine oxidase molecules in the body and decreases the amount of uric acid formed, thereby preventing uric acid crystals from forming and collecting in the joints of affected individuals.

Problem 3–19

A protein that transports amino acids across the cell membrane was found to bind only a few amino acids efficiently. To find the specificity, many different amino acids and structural analogs were used as competitive inhibitors in transport studies, with radioactive [^{14}C]histidine as the substrate. The v_0 values were reported as μmol [^{14}C]his transported into 10^6 cells/min. The K_i values calculated from Lineweaver–Burk plots are shown in Table 3–20. Comparing the structures of histidine and the competitive inhibitors, what can you conclude about the characteristics of molecules that this protein binds at its active site? (For structures of amino acids, see Chapter 2.)

Table 3–20. K_i Values[a]

Amino Acid or Analog	K_i ($\times 10^{-6}$ M)	Structure
L-lys	2	
L-arg	3	
L-gly	285	histamine
L-asp	450	
D-his	340	
Histamine	390	dehydrourocanate
Dihydrourocanate	285	
D-arg	355	

[a] All the assays were performed at pH 5.9. K_m for L-histidine was found to be 10 μM when the pH of the binding assay was maintained at 5.9. When the assays were performed at pH 6.9, K_m for L-histidine was found to be 190×10^{-6} M.

Solution:

The protein binds amino acids in the L form with a positive charge on the R group. The protein prefers the L form, as shown by the difference between K_i values of L-arginine and D-arginine. L-arg is an excellent competitive inhibitor, but D arg is very poor. Also, D-histidine competes poorly with [^{14}C]L-his for binding.

From these data it appears that only amino acids with a positively charged R group are good inhibitors. When the pH is raised so that L-his loses the proton from its R group, the K_m rises from 10 μM at pH 5.9 to 190 μM at pH 6.9.

Histamine differs from histidine only in the lack of an α-COO$^-$ group. The high K_i value for histamine indicates that it does not bind efficiently to the protein. Dihydrourocanate differs from histidine in lack of an α-NH$_3$ group. The high K_i value for dihydrourocanate indicates that it does not bind to the protein either. From these observations we may conclude that this protein binds only amino acids with α-NH$_3^+$ groups and α-COO$^-$ groups.

Problem 3–20

The enzyme monoamine oxidase is found in most cells and catalyzes removal of the amino group from compounds such as tyramine. In some people, inhibition of this enzyme prevents severe depression. The specificity of this enzyme has been determined in order to develop more effective drugs which will act as competitive inhibitors. The K_m values for natural substrates are shown below and also the K_i values for other compounds used as competitive inhibitors of tyramine.

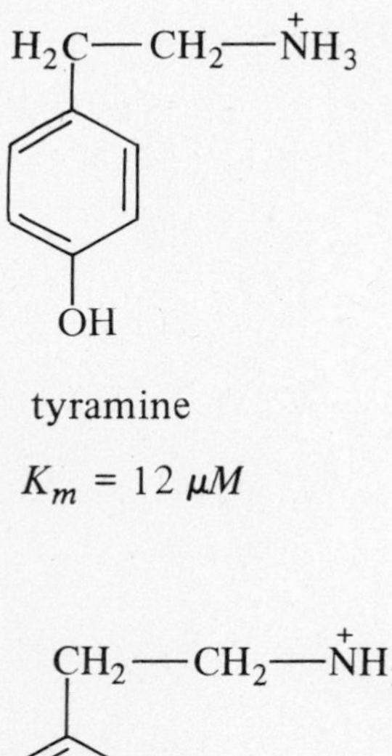

tyramine

$K_m = 12 \ \mu M$

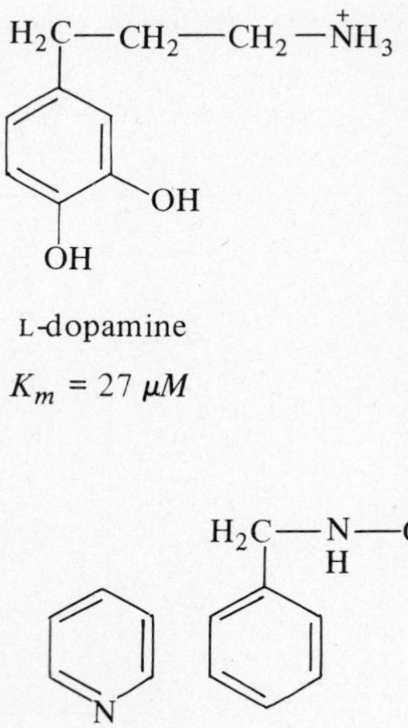

L-dopamine

$K_m = 27 \ \mu M$

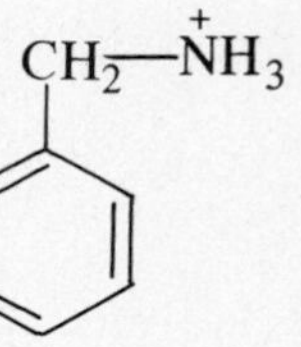

benzylamine

$K_m = 70 \ \mu M$

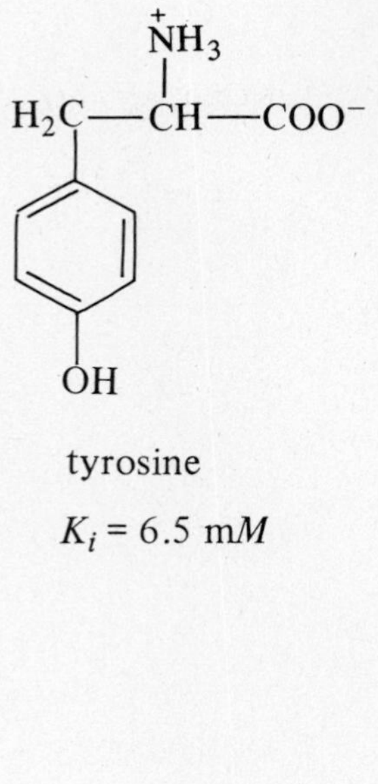

tyrosine

$K_i = 6.5 \ mM$

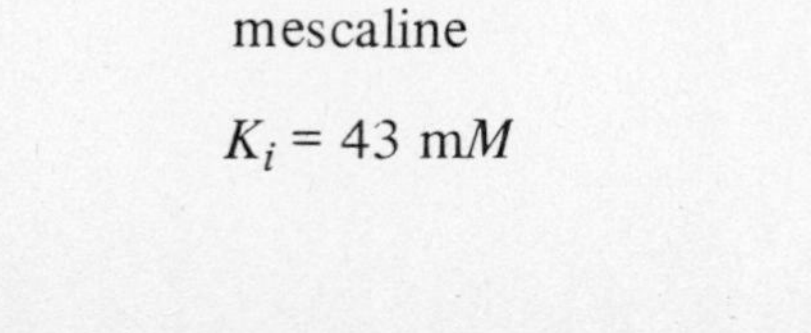

mescaline

$K_i = 43 \ mM$

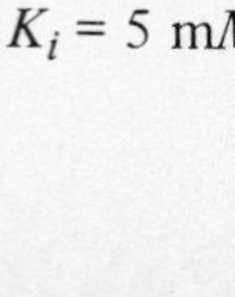

pyridine

$K_i = 5 \ mM$

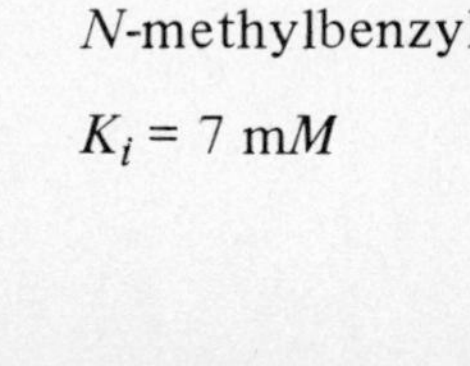

N-methylbenzylamine

$K_i = 7 \ mM$

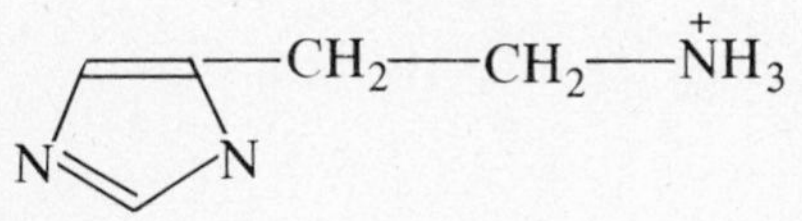

$$CH_2-CH_2-\overset{+}{N}H_3 \qquad CH_3-CH_2-\overset{+}{N}H_3 \qquad O=\overset{|}{C}-NH-\overset{+}{N}H_3 \qquad H_2C-COO^-$$

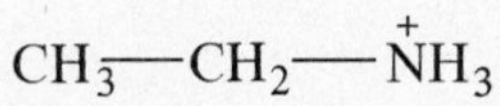

histamine	ethanolamine	isoniazid	homogentisic acid
$K_i = 2$ mM	$K_m = 5$ mM	$K_i = 50$ μM	$K_i = 6$ mM

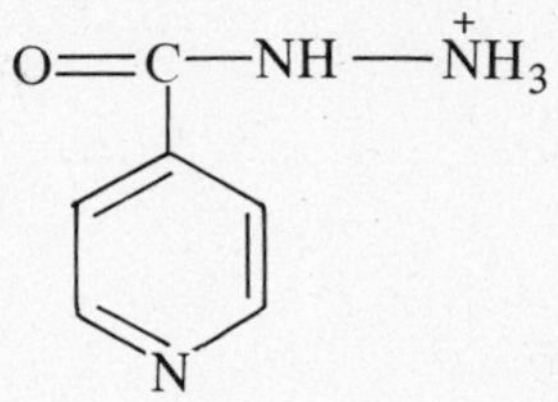

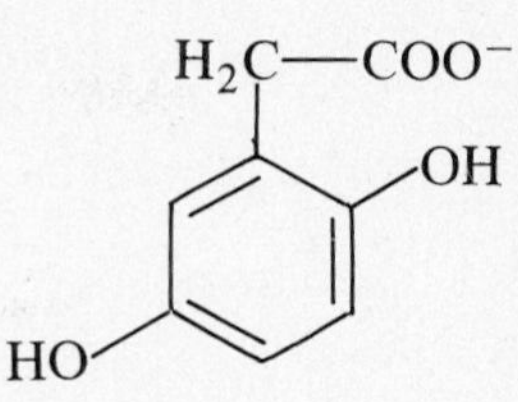

Which combination of the following groups are *essential* on a compound for it to bind to the active site of this enzyme?

    (a) Pyridine ring or benzene ring   (b) Primary carboxyl group

    (c) Primary amine group   (d) Net negative charge

    (e) OH group on ring

Solution:

By comparing the K_m and K_i values we may conclude that all compounds which bind effectively to the enzyme have (c) *a primary amine*. For example, compare the K_i values for benzylamine and N-methylbenzylamine. Since histamine and ethanolamine bind poorly, the substrate apparently must have (a) *pyridine or a benzene ring* as well as a primary amine group to bind. While the enzyme has a slightly higher affinity for tyramine than for benzylamine, an OH group on the ring is not essential for binding, as shown by the relatively low K_i values for benzylamine and isoniazid.

Determining the Effective [S] When the Substrate Exists in Different Forms

Substrates for enzymes often exist in cells in more than one form, but many enzymes only bind one form. For example, there are D and L forms of molecules, α and β anomers of sugars, and protonated or unprotonated forms of chemical groups found on biomolecules.

Weak Acids as Substrates

Many substrates are weak acids that may exist in protonated and unprotonated forms, depending on the pH of the solution. In some cases, enzymes bind only one form, and therefore to determine the correct [S], the concentration of the protonated or unprotonated form must be determined as in the following problem.

Problem 3–21

The enzyme histidine deaminase binds histidine and cleaves the amino group, producing NH_3 and dihydrourocanate.

$$\text{histidine} \longrightarrow \text{dihydrourocanate} + NH_3$$

Assuming that this enzyme binds only histidine molecules with a net charge of +1, calculate the v_0 of this reaction given that [histidine] = 30 μM ;   pH of reaction mixture = 6.7

K_m = 25 μM ; V_{max} = 3.0 μmol/min/mg protein ;   0.1 mg protein is present

Solution:

Histidine may exist in the following forms:

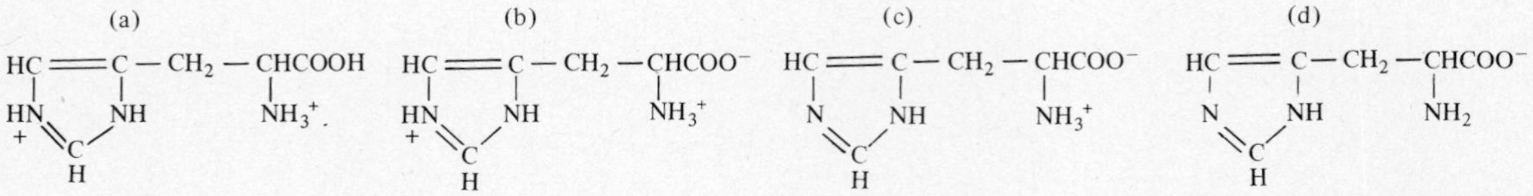

Because the enzyme only binds histidine molecules that have a net charge of +1, the effective [S] is less than the total histidine concentration. At pH 6.7, histidine will exist as form (b) (weak acid) or (c) (its conjugate base). Form (b) is the only one with a net charge of +1. The Henderson–Hasselbalch equation is used to determine the concentration of this form of histidine at pH 6.7. (The pK of the imidazole group is 6.0.)

$$\text{pH} = \text{p}K + \log \frac{[\text{base}]}{[\text{acid}]}$$

where [base] = [A^-] = [his^0] and [acid] = [HA] = [his$^+$]. Let [his$^+$] = x; then [his^0] = 30 $\mu M - x$. Then

$$6.7 = 6.0 + \log \frac{30 - x}{x} ; \qquad 0.7 = \log \frac{30 - x}{x}$$

antilog(0.7) = 5

Therefore,

$$\frac{30 - x}{x} = 5 ; \quad 30 - x = 5x ; \quad 6x = 30 ; \quad x = 5 \ \mu M$$

[Histidine] with a net charge of +1 is 5 μM.

   Alternative solution:

$$6.7 = 6.0 + \log \frac{[\text{base}]}{[\text{acid}]} ; \quad \log \frac{[\text{base}]}{[\text{acid}]} = 0.7$$

$$\text{antilog}(0.7) = 5 ; \quad \frac{[\text{base}]}{[\text{acid}]} = \frac{5}{1}$$

Since [base] + [acid] = 30 μM:

$$5x + 1x = 30 \ \mu M$$

$$x = 5 \ \mu M = [\text{acid}] = [\text{his}^+]$$

From the calculations above, the effective [S] is 5 μM. To determine the v_0 in this problem, the Michaelis–Menten equation is used:

$$v_0 = \frac{V_{max}[S]}{K_m + [S]}$$

$$= \frac{3 \ \mu\text{mol/min/mg} \times 5 \ \mu M}{25 \ \mu M + 5 \ \mu M} = 0.5 \ \mu\text{mol/min/0.1 mg enzyme}$$

$$= 5.0 \ \mu\text{mol/min/mg}$$

Problem 3–22

The enzyme aconitase is found in mitochondria and catalyzes the conversion of citrate to isocitrate. Assuming that the enzyme aconitase binds only unprotonated citrate molecules, calculate the amount of isocitrate formed under the following conditions:

$$V_{max} = 5.5 \ \mu\text{mol/min/mg enzyme}; \quad K_m = 0.075 \ \text{m}M$$

$$[\text{citrate}] = 200 \ \mu M; \quad \text{pH} = 6.4$$

The pK values for the carboxyl groups on citrate are 3.1, 4.7, and 6.4.

Solution:

To determine the effective [S] at pH 6.9, the Henderson–Hasselbalch equation must be used. Since the pH of the reaction is well above the pK values for two of the three carboxyl groups, we may safely assume that at this pH more than 99% of the citrate molecules exists as cit^{2-} or cit^{3-}

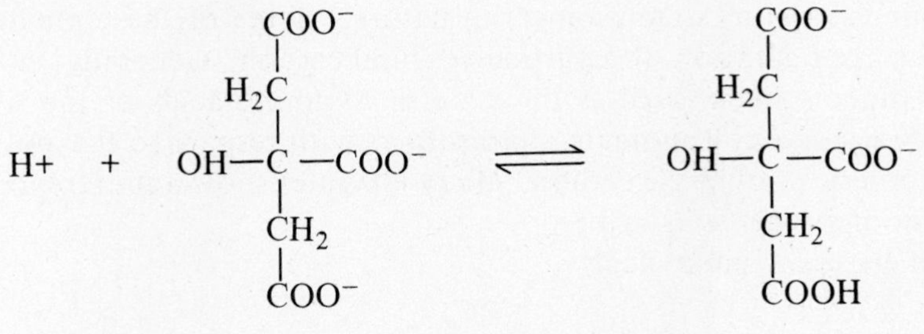

The pK of the third carboxyl group at 6.4 is the one to consider in determining the concentration of unprotonated citrate, cit^{3-}.

$$pH = pK + \log \frac{[\text{conjugate base}]}{[\text{conjugate acid}]}$$

$$6.9 = 6.4 + \log \frac{[cit^{3-}]}{[cit^{2-}]}$$

Let $[cit^{3-}]$ be x. Since the concentration of all forms of citrate is 200 μM, $[cit^{2-}]$ must be $200 - x$:

$$6.9 = 6.4 + \log \frac{x}{200 - x}$$

$$0.5 = \log \frac{x}{200 - x}$$

The antilog of 0.5 is 3.16. Therefore,

$$3.16 = \frac{x}{200 - x}$$

$$632 - 3.16x = x$$

$$632 = 4.16x$$

$$x = 152.2$$

Thus $[cit^{3-}] = 152 \mu M$. Therefore, the effective $[S] = 152 \mu M$ citrate. Thus us

$$v_0 = \frac{5.5 \times 152}{75 + 152} = \frac{837}{227} = 3.68 \ \mu mol/min/mg \ \text{enzyme}$$

Stereoisomers as Substrates

Many compounds exist as stereoisomers in nature. Some of these are designated D or L depending on the configuration at a particular chiral carbon. Generally, enzymes bind only one of the two stereoisomers, such as the L form of amino acids or the D form of sugars. Reducing sugars exist as α or β anomers, steroisomers with respect to the position of the OH group on the anomeric (carbonyl) carbon. Many enzymes show a preference for binding the α or the β anomer of a sugar substrate.

D,L stereoisomers of an amino acid:

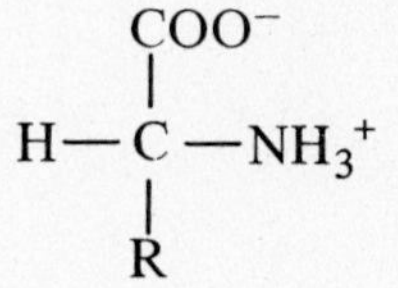

$$COO^-$$
$$|$$
$$H—C—NH_3^+$$
$$|$$
$$R$$

D form

$$COO^-$$
$$|$$
$$H_3\overset{+}{N}—C—H$$
$$|$$
$$R$$

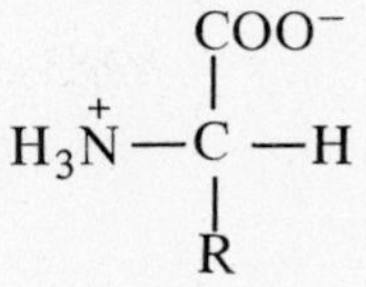

L form

α,β anomers of the reducing sugar, glucose:

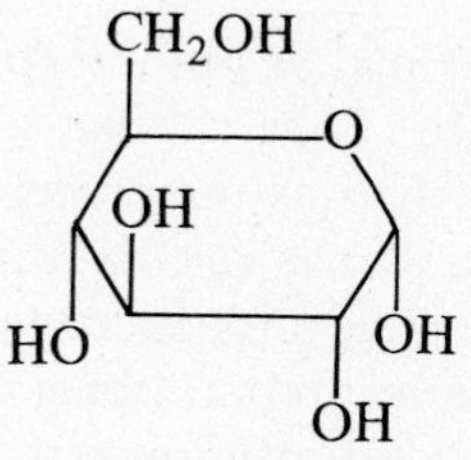

CH₂OH
(structure)
OH
HO
OH
OH

α anomer, D-glucose

CH₂OH
(structure)
OH
OH
HO
OH

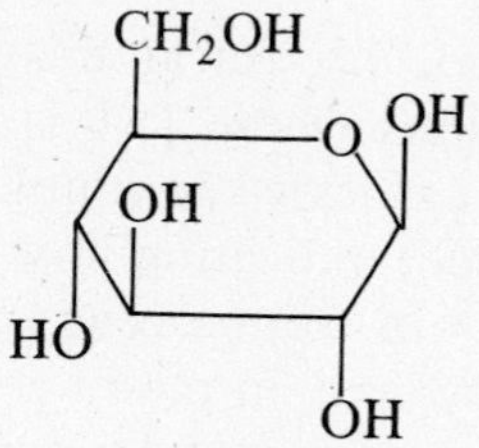

β anomer, D-glucose

Problem 3-23

The enzyme hexokinase binds two different substrates, α-D-glucose and ATP, and catalyzes the production of two products, glucose-6-PO$_4$ and ADP. Assume that $V_{max} = 5$ μmol/min/mg protein, and the K_m value for D-glucose is 50 μM and for ATP is 38.0 μM. [ATP] = 2.0 mM and [glucose] (all forms) is 75.0 μM. Calculate the amount of glucose-6-PO$_4$ that is formed in 1.0 min. Only 34% of glucose in solution is in the α-D form; 66% exists as the β anomer of D-glucose.

Solution:

The substrate at saturation levels is ATP since [ATP] is greater than 50 times its K_m. Therefore, the v_0 of this reaction will not depend on [ATP] but will be a function of [D-glucose] since it is close to its K_m.

[α-D-glucose] is 34% of 75 μM = 25.5 μM.

The K_m for α-D-glucose = 50 μM.

Under these conditions, the Michaelis–Menten equation may be used to determine initial velocity,

$$v_0 = \frac{V_{max}[S]}{K_m + [S]}; \qquad [S] = 25.5 \ \mu M \quad \text{and} \quad K_m = 50 \ \mu M$$

$$= \frac{5 \ \mu\text{mol/min/mg} \times 25.5 \ \mu M}{50 \ \mu M + 25.5 \ \mu M}$$

$$= \frac{127.5}{75.5}$$

$$= 1.69 \ \mu mol/min/mg$$

Then 1.69 μmol glucose-6-PO$_4$ is formed per minute per milligram of enzyme.

Turnover Number (k_{cat})

The turnover number or k_{cat} for an enzyme is defined as μmol P formed/min/μmol enzyme under optimal conditions where [S] is very high. It is similar to V_{max} except that the amount of enzyme is expressed in μmoles rather than as mg protein. It is a measure of the efficiency of an enzyme, indicating how much substrate can be turned over (i.e., converted to product) in 1 min, but can be used only when the molecular weight of the enzyme has been determined. However, in many crude enzyme preparations, the molecular weight of the enzyme being assayed is not known, so the turnover number cannot be determined. Turnover numbers for some common enzymes are listed in Table 3–21.

Table 3–21. Turnover Number (k_{cat}) for Some Common Enzymes

Enzyme and Reaction Catalyzed	k_{cat}
Catalase $2H_2O_2 \rightarrow 2H_2O + O_2$	10^7
Carbonic anhydrase $H_2CO_3 \rightarrow H_2O + CO_2$	10^6
Alcohol dehydrogenase ethanol + NAD$^+$ $\rightarrow$ acetaldehyde + NAD$\cdot$H + H$^+$	10^3
Transaminase (SGOT) glutamate + oxaloacetate $\rightarrow$ ketoglutarate + asp	10^3

Problem 3–24

A penicillin-resistant mutant strain of bacteria produces an enzyme (penicillinase) that degrades penicillin. Assuming that penicillinase has a k_{cat} of 2 X 10^3, how many minutes would it take 10 nmol of this enzyme to degrade 250 mg of penicillin? (Assume that MW = 350.)

Solution:

k_{cat} = 2 X 10^3 μmol product formed/min/μmol enzyme

k_{cat} may also be expressed as

$$k_{cat} = \frac{2 \times 10^3 \ \mu mol \ substrate \ consumed}{min \cdot \mu mol \ enzyme} \times \frac{1 \ \mu mol \ enzyme}{1 \times 10^3 \ nmol} = \frac{2 \ \mu mol \ S}{min \cdot nmol \ enzyme}$$

The amount of penicillin present is 250/350 mg = 0.71 mmol or 710 μmol. If 1 nmol of enzyme consumes 2 μmol S/min, then 10 nmol enzyme consumes 20 μmol/min. Since 10 nmol of penicillinase degrades 20 μmol of penicillin in 1.0 min, it would take 710/20 = 35.5 min to degrade 250 mg of penicillin.

$$\frac{710 \ \mu mol S}{20 \ \mu mol S} = \frac{x \ min}{1 \ min}; x = 35.5 \ min$$

Problem 3–25

The enzyme alcohol dehydrogenase catalyzes the oxidation of ethanol to acetaldehyde. Assuming that the turnover number for this enzyme is 1.1×10^3/min and its MW = 10^5, how much ethanol can be degraded to acetaldehyde by 1.5 mg of this enzyme in 60 min?

Solution

Since k_{cat} = 1.1×10^3, then 1 μmol of alcohol dehydrogenase catalyzes formation of 1.1×10^3 μmol acetaldehyde in 1.0 min.

$$\mu mol \ enzyme = \mu g \ enzyme/MW \ in \ \mu g = \frac{1500}{10^5} = 0.015 \ \mu mol$$

Then 0.015 μmol of the enzyme forms $1.1 \times 10^3 \times 0.015$ = 16.5 μmol of acetaldehyde per minute, so 16.5 μmol of ethanol will be degraded in 1.0 min. In 60 minutes, 60 $\times$ 16.5 = 990 μmol of ethanol is degraded.

Enzymes That Deviate from Michaelis–Menten Kinetics

The enzymes that regulate metabolic pathways in cells usually consist of several polypeptides (subunits) held together largely by noncovalent forces (electrostatic attractions, hydrophobic interactions). These enzymes are termed *multimeric*. Each subunit (monomer) is usually required for enzyme activity even though some of the subunits may not contain an active site. With some (but not all) multimeric enzymes the phenomena of *cooperativity* and *allosteric inhibition* are seen. In both cases, the subunits communicate by slight conformational changes when a substrate molecule or an allosteric inhibitor is bound to one of the subunits.

Cooperativity

Cooperativity occurs when the *binding of a substrate molecule to one active site changes the affinity for the substrate binding to another active site.* The binding of a substrate molecule to one active site may change slightly the overall conformation of a protein, altering the other active sites. When an active site is changed even slightly, its affinity for the substrate molecule may change. In *positive cooperativity*, the binding of the first substrate molecule to the enzyme increases the enzyme's affinity for the remaining substrates.

The Michaelis–Menten model assumes that the binding of a substrate molecule to one active site does not affect binding at other active sites. Therefore, enzymes showing co-operativity will *not* have the same curves for v_0 versus [S] plots or the straight lines for $1/v_0$ versus $1/$[S] plots as enzymes showing no cooperativity. The kinetic data from an enzyme exhibiting positive cooperativity are shown in Fig. 3–17a. Although the time course data plots (not shown) appear to be the same as for other enzymes, *the v_0 versus [S] plot is sigmoidal* or S shaped (Fig. 3–17a) instead of the rectangular hyperbola seen in Fig. 3–17b for an enzyme showing no cooperativity. In comparing Fig. 3–17a and b, note how much *lower* the v_0 is at 10 μM [S] in Fig. 3–17a (positive cooperativity), relative to the v_0 at 10 μM [S] in Fig. 3–17b, despite similar V_{max} values for both enzymes.

Enzymes with positive cooperativity are able to respond rapidly to changes in [S] over a very narrow range around $K_{0.5}$, which is [S] at $1/2$ V_{max}. Generally, this narrow range corresponds to the physiological [S] in the cell.

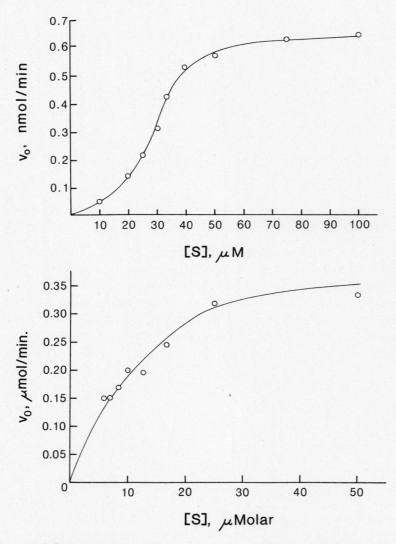

Figure 3–17. v_0 versus [S] plots for two different enzymes: (a) the enzyme is composed of several poly-peptides and positive cooperativity in substrate binding is demonstrated by the sigmoid (S shaped) curve; (b) a different enzyme which contains four subunits, each with one active site, demonstrates the rectangular hyperbola characteristic of Michaelis–Menten kinetics.

Allosteric Regulation of Enzyme Activity

Both cooperativity in substrate binding and allosteric inhibition are the result of conformational changes in the enzyme caused by the binding of a ligand, either substrate or allosteric effector. Often, multimeric enzymes that exhibit cooperativity in substrate binding also show allosteric inhibition, that is, binding of an inhibitor molecule at an *allosteric site*. In some enzymes the active sites are found on the catalytic subunits and the allosteric sites are found on *regulatory subunits*.

Problem 3-26

ATCase (aspartate transcarbamylase) is the first enzyme in the pathway that synthesizes pyrimidine nucleotides in cells. ATCase catalyzes the formation of carbamoyl aspartate from aspartate and carbamoyl phosphate.

carbamoyl phosphate   aspartate   carbamoyl L-aspartate   phosphate

From the data in Table 3–22, determine:

(a) Whether there is cooperativity in the binding of aspartate to ATCase.

(b) How the compound CTP affects the reaction.

(c) What the approximate concentration of aspartate is in the cell.

Table 3–22. Initial Velocities Obtained from Time Course Data (Not Shown)

[S] (mM)	v_0, μmol/min	v_0 (+ CTP), μmol/min
1	6.7	1.5
2	35.6	3.8
3	40.5	12.5
4	54.2	14.9
5	72.9	21.3
6	92.0	37.8
7	99.8	48.4
8	104.9	55.3
9	115.8	71.8
10	117.3	83.3
15	128.4	104.2
20	128.4	126.7

Source: Unpublished data from James R. Wild.

Assay conditions: 40 mM KH_2PO_4 (buffer); 5.0 mM carbamyl PO_4 (other substrate at saturation levels 25 × K_m).

Solution:

(a) The data plotted in Fig. 3–18 for v_0 versus [S] with no CTP show a slightly sigmoidal curve typical of enzymes with positive cooperativity, rather than the rectangular hyperbolas seen with enzymes exhibiting Michaelis–Menten kinetics.

(b) The CTP inhibits the v_0 at each [S] but much less at higher [S]. CTP is known to bind to the *allosteric* site rather than the active site.

(c) The approximate concentration range for aspartate in the cell corresponds to the narrow range of maximum rate of enzyme response around $K_{0.5}$. $K_{0.5}$ is the [S] at $1/2$ V_{max}.

From Fig. 3–18, V_{max} is *estimated* to be about 130 nmol/min. The [aspartate] at $v_0 = 65$ nmol/min is approximately 4 mM. The concentration range that corresponds to the maximum response appears to be between 3 and 6 mM aspartate. Therefore, [asp] in the cell is probably around 3 to 6 mM.

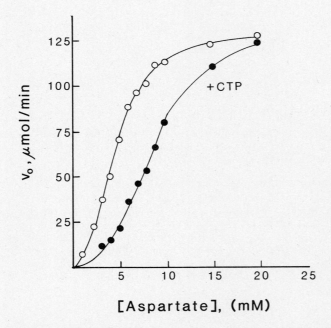

Figure 3–18. Initial velocities at varying [asp] in the presence and absence of the allosteric inhibitor, CTP.

Exercises

The answers are provided at the back of the book.

3-1. How much malate may be formed by 1.0 mg of the enzyme fumarase in 1.0 min under the following assay conditions?

[fumarate] = 80 μM; K_m for fumarate = 90 μM; V_{max} = 0.6 μmol/min/mg enzyme

Fumarase catalyzes the reaction

$$\text{fumarate} + H_2O \longrightarrow \text{malate}$$

3-2. Benzopyrene is a carcinogen found in cigarette smoke. The enzyme aryl hydrocarbon hydroxylase (AHH-P450 reductase) catalyzes the addition of a hydroxyl group to give 7-hydroxybenzopyrene in lung tissue. Unfortunately, the 7-OH product appears to be a more potent carcinogen than the benzopyrene substrate.

$$\text{NADPH} + \text{benzopyrene} \longrightarrow \text{NADP}^+ + \text{7-hydroxybenzopyrene}$$

Initial velocities were determined as μmol product formed/min/0.2 mg enzyme at different [S] in the presence and absence of a structural analog of benzopyrene. From the data below, determine:

(a) K_m (b) V_{max}

(c) The type of inhibition   (d) K_i

[benzopyrene] (μM)	v_0 (μmol/min/0.2 mg)	
	No Inhibitor Present	[Inhibitor] = 1.28 μM
2	0.16	0.125
4	0.25	0.20
5	0.26	0.22
10	0.38	0.31
50	0.495	0.45

3-3. The cofactor NAD$^+$ is formed in the cell by a reaction that adds an adenosine monophosphate to nicotinamide mononucleotide.

$$\text{NMN} + \text{ATP} \xrightarrow{\text{adenyl transferase}} \text{NAD}^+ + \text{P-P}$$

The reaction is catalyzed by the enzyme adenyl transferase. Calculate the amount of NAD$^+$ that would be formed given the following conditions.

[NMN] = 30 μM [ATP] = 2 mM V_{max} = 75 nmol/min/mg

K_m(NMN) = 50 μM K_m(ATP) = 2 μM

3-4.   The following time course data were obtained for an enzyme-catalyzed reaction in the presence and absence of an inhibitor. 20 μg enzyme was present in each assay.

(a) Calculate the K_m for this substrate, the V_{max}, and the K_i for the inhibitor.

(b) What type of inhibition is caused by this inhibitor?

| | nmol Product Formed | | | | | |
| | No Inhibitor | | | [Inhibitor] = 25 μM | | |
[S] (μM)	30 sec	60 sec	90 sec	30 sec	60 sec	90 sec
50	4.3	8.3	16.5	2.5	5.1	10.3
25	3.5	6.6	13.0	1.7	3.5	6.8
20	3.1	5.9	12.0	1.5	3.1	5.9
10	2.0	3.9	7.3	0.9	1.8	3.5

3-5.   The enzyme glucose oxidase catalyzes the oxidation of glucose to gluconic acid. This enzyme has been isolated from some fungi and is used commercially to test for the presence of glucose in urine. From the data given below, determine the specificity of this enzyme for its sugar substrate.

The K_m for β-D-glucose was determined to be 31 μM. Each of the following sugars were used as potential competitive inhibitors of β-D-glucose.

Potential Inhibitor	K_i (μM)	Potential Inhibitor	K_i (μM)
α-D-glucose	250	β-D-glucosamine	37
β-D-galactose	360	β-D-fructose	390
β-D-ribose	390	α-D-glucosamine	290
α, β-L-glucose	400		

Which of the following correspond to a structural feature of sugars that is required for the enzyme to bind it?

(a) α anomer   (b) β anomer

(c) D-stereoisomer   (d) L stereoisomer

(e) Hexose   (f) Pentose

(g) Aldose   (h) Ketose

3-6.   The turnover number for the enzyme β-galactosidase is approximately 12,500. β-Galactosidase cleaves the disaccharide lactose to yield glucose and galactose. How much lactose can 5 nmol of this enzyme cleave in 10 min?

3-7.   Histidine decarboxylase binds histidine and cleaves the carboxyl group, releasing CO_2 and histamine. Assuming that this enzyme binds only histidine molecules with a net charge of zero, calculate the rate of histamine formation under the following conditions:

pH = 6.5   [histidine] = 30 μM K_m for his = 25 μM V_{max} = 3 μmol/min/mg protein

0.1 mg of protein was used.

Chapter 4

ISOLATION, PURIFICATION, AND ANALYSIS OF PROTEINS AND NUCLEIC ACIDS

In its present state, our technology does not allow us to study the properties of proteins and nucleic acids in the intact cell. Since these molecules are synthesized only in cells, the cells must be disrupted in order to extract them. The basic methods of purification and analysis of proteins and nucleic acids are quite similar. To separate proteins and nucleic acids from the other components of the cell and from each other, chemical methods are used which take advantage of their different solubilities in high salt solutions and in alcohols. Physical methods such as centrifugation, gel exclusion chromatography, and electrophoresis are used to further purify and to characterize particular proteins or nucleic acids.

Many of the steps used in protein purification are performed to maintain the protein in its native active state. Understanding the structure of proteins will increase your appreciation for the various techniques to be discussed later in this chapter.

Structure of Proteins and Nucleic Acids

Proteins

Proteins are linear polymers of amino acids (primary structure) held together by covalent peptides bonds. The linear polymer normally assumes a particular secondary structure depending on the amino acid sequence. Some amino acid sequences naturally form an α-helix, while other sequences form a β-pleated sheet or take a coil form. The α-helix, β-pleated sheet, and coil parts of each protein are folded into a tightly wound "ball of yarn" (tertiary structure) to form globular proteins, many of which function as enzymes in the cell.

Covalent bonds are important in holding the amino acid residues together in primary structure, but the most important forces in maintaining secondary and tertiary structure are noncovalent forces. Hydrogen bonding between atoms in different peptide bonds maintains the α-helical parts of proteins. Electrostatic attractions, hydrophobic interactions, and hydrogen bonds maintain the protein's tertiary structure. Covalent disulfide bonds between cysteine residues also help stabilize tertiary structure in some proteins.

Many proteins in cells have subunit structure (quaternary structure). Each subunit is a complete polypeptide with tertiary structure, and together the subunits form a functional protein that may catalyze a specific reaction. The main forces holding subunits together are electrostatic attractions, hydrogen bonds, and hydrophobic interactions. The overall geometry of the subunits is also important in their fitting together.

During the purification process, great care must be taken to maintain these forces and prevent the protein from losing its quaternary, tertiary, or secondary structure. One factor that can interfere with H bonds and hydrophobic interactions is an increase in temperature. Most proteins unfold at temperatures above 30°C, while most are quite stable between 4 and 15°C. Therefore, during the purification process, the protein sample is kept in an ice bath to prevent temperature increases. A change in pH may affect electrostatic interactions and hydrogen bonding, which are essential for the proper folding of the protein, and therefore a buffer is used to prevent pH changes. Free sulfhydryl groups (−SH) are essential for the catalytic activity of many proteins. When these proteins are being purified, antioxidants such as dithiothreitol must be present to prevent oxidation of these sensitive groups.

DNA Structure

All nucleic acids are polymers of nucleotide monophosphates. DNA differs from RNA in its sugar, its pyrimidine bases, and its double-stranded helical nature. All DNA molecules contain the nucleoside monophosphates, dAMP, dGMP, dCMP, and dTMP linked in specific sequence by 3′, 5′-phosphoester bonds as shown below.

The amino acid sequence in a protein is dictated by the sequence of bases in the DNA of the gene that codes for it. The double-stranded DNA helix carries the code for a particular protein on one strand, the *sense strand*. The complementary strand is called the *antisense strand*. The sense strand is transcribed by RNA polymerase, and the RNA that is formed is complementary to the sense strand.

In general, all strands of RNA and DNA are synthesized from a DNA template and always in the direction opposite to that of the DNA template. The direction of synthesis for both DNA and RNA is 5′ to 3′. The nucleotide that begins every newly forming strand has a free 5′-triphosphate while its 3′-OH is joined in a phosphoester linkage to the 5′-PO$_4$ of the second nucleotide. In the example below, the DNA is being transcribed and the resulting mRNA is of opposite polarity to the sense strand of DNA from which it was copied.

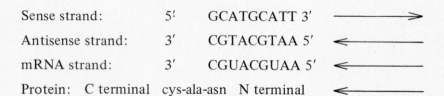

Sense strand:	5′	GCATGCATT 3′	⟶
Antisense strand:	3′	CGTACGTAA 5′	⟵
mRNA strand:	3′	CGUACGUAA 5′	⟵
Protein:	C terminal	cys-ala-asn N terminal	⟵

The arrows indicate the direction of synthesis for each molecule. *Translation* is the term used to describe protein synthesis where the base sequence in mRNA is translated into the amino acid sequence in a protein.

The amino acid sequence for which this mRNA codes may be determined from the codon dictionary (Fig. 4–10, p. 172). The N-terminal aminoa cid is coded for by a codon at the 5′ end of the message, in this case by the codon, AAU. The direction of reading the message is *always* 5′ end to 3′ end, so that the N-terminal amino acid is the first one in the growing protein chain and the C-terminal amino acid is the last to be added.

Isolation and Characterization of Proteins and Nucleic Acids

The basic techniques used to purify and analyze proteins and nucleic acids are often more refined versions of the techniques used to isolate these macromolecules from cell extracts. For example, centifugation is used to remove cell fragments and unbroken cells from the soluble cell contents, while ultracentrifugation is used to separate proteins from DNA or RNA from DNA by spinning the solutions at extremely high speeds. In the following section, a few basic techniques are described briefly, but the reader is referred to other books for detailed discussions of methods and procedures.

Isolation of Proteins and Nucleic Acids from Cell Extracts

Proteins and nucleic acids are synthesized in cells where they function to direct the cell's metabolism. To isolate, purify, and study these molecules, the cells must be ruptured and the proteins or nucleic acids separated from the other cell contents.

Disruption of Cells. Many types of cells, especially fungal and plant, require powerful tools to break the tough cell walls. In these instances, "power tools," such as a Waring blender or a hydraulic press (French press), are used. Tissues with cells that lack cell walls, such as mammalian cells, may be disrupted by gentler methods, including osmotic manipulations of the surrounding fluid, sonication (bombarding by sound waves), and manually homogenizing in a glass homogenizer.

Since most of these methods produce heat, the solutions containing the cells to be broken are kept in an ice bath to prevent large temperature increases which may denature proteins. It is essential that a constant pH be maintained throughout the purification procedure by using buffered solutions.

Centrifugation. After the cells are broken, the contents are collected in a buffer and centrifuged to separate the insoluble cell fragments from the soluble cell contents. The pellet contains the insoluble cell fragments and unbroken cells, while the supernatant contains the soluble macromolecules, such as most proteins and nucleic acids. The first step in separating the proteins from nucleic acids is usually salting out of proteins. In this procedure, the proteins precipitate out of the high salt solution and the DNA and RNA remain dissolved.

Salting Out Proteins. Proteins exist in cells in a hydrated state, that is, water molecules coat the hydrophilic residues on the outer surface of the protein, allowing it to remain in solution in the cell. Other charged molecules and ions are also coated with water molecules. When the concentration of salt ions increases to very high levels [e.g., $1.0\ M$ or 30 to 50% $(NH_4)_2SO_4$], many of the water molecules leave the protein's surface and coat the salt ions. Unhydrated protein molecules prefer to associate with each other and as a result, precipitate from solution. This phenomenon is called *salting out*.

Salting out of proteins is an effective step in most protein purifications, allowing proteins to be separated from nucleic acids, some other proteins, and many other cellular molecules. Different proteins precipitate at different salt concentrations. Some proteins are insoluble in 30% ammonium sulfate (AS) solutions, but many others are soluble in solutions up to 60%. By stepwise additions of AS to solutions, proteins may be separated from each other without destroying the protein's structure or activity.

DNA Precipitation. DNA is soluble in solutions of high ionic strength (high salt concentration) but generally insoluble in alcohols. After most of the proteins have been precipitated in high salt solutions, cold alcohol is added slowly and DNA precipitates at the interface. If the DNA concentration is high, the long, fibrous DNA may be wound out of solution on a glass rod.

Pruification and Analysis of Proteins and Nucleic Acids

Proteins and nucleic acids that have been isolated from cell extracts as described above may be purified and analyzed using the techniques described below.

Note: The techniques of affinity chromatography and ultracentrifugation are basic tools used to purify proteins and nucleic acids. The reader is referred to other books for descriptions of these techniques.

Gel Exclusion Chromotography. The relative size of a molecule may be determined using gel exclusion chromatography. With this technique a column is filled with tiny beads suspended in buffer (Fig. 4–1 and Table 4–1). The microscopic beads are made from cross-linked polymers of glucose or acrylamide, much like a practice golf ball or whiffle ball.

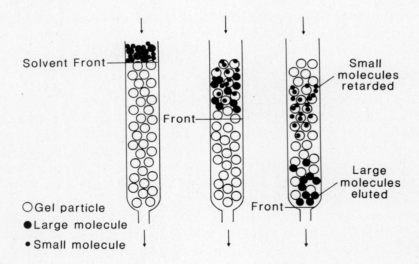

Figure 4–1.

The protein is dissolved in a buffer and is added to the top of the column bed. Buffer is added to the column and the protein is carried through the bed at a rate depending on the size of the protein and the size of the pores in the beads. If the pores are very small (as in Sephadex G-10 beads) the protein will be unable to enter the pores and will flow through the column very quickly (Fig. 4–1), eluting at the end of the void volume. However, if the pores are relatively large, the protein will enter the beads and will slowly flow through the column as more buffer is added.

Note: The void volume is the volume of the beads in the column bed.

The fluid that comes off the column is called the *eluent* and is caught in successive fractions of one or more milliliters (Fig. 4–2). The larger proteins elute (come off the column) first and are found in the first few fractions after the void volume. Smaller proteins elute much later because they are slowed by entering the beads. Proteins that are too large to enter the beads elute at the end of the void volume, and very little may be determined about their sizes except that they exceed the exclusion limits of the beads.

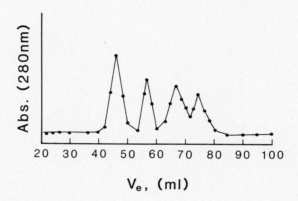

Figure 4–2. Elution of proteins from Sephadex G-50 column. Albumin appears in fractions 42–50, carboxypeptidase in fractions 54–60, myoglobin in fractions 62–72 and cytochrome *c* in fractions 72–80.

The proteins in each fraction may be detected by measuring the absorbance of the fraction at 280 nm in a spectrophotometer (see Chapter 1). The absorbances of each fraction may be plotted to give an elution profile as shown in Fig. 4–2. Many experiments have demonstrated that the molecular weight (M_r) of a spherical molecule is inversely related to the volume of eluent (V_e) required to elute it from a gel exclusion column. In fact, the (V_e) at the peak fraction for a protein is inversely proportional to the log of its molecular weight (M_r). Since most proteins are not spherical, this method allows only an approximation of the M_r of a protein.

Problem 4–1

A solution of the proteins, carboxypeptidase, cytochrome *c*, myoglobin, and albumin was placed on a column containing a bed of Sephadex G-50 beads, which have pores to exclude molecules larger than about 40,000 molecular weight. Buffer was added to elute the proteins and the elution profile shown in Fig. 4–2 was obtained. What are the relative sizes of the proteins in the mixture?

Solution

Albumin is the largest since it eluted first. Carboxypeptidase is larger than myoglobin, which is larger than cytochrome *c*, which eluted last.

Table 4-1. Characteristics of gel exclusion beads

Type of Bead	Molecular Weight of Molecules Excluded	Bead Size (μm)	Bead Composition
Sephadex G-10	1,000	100	Cross-linked dextrans (glucose polymers)
Sephadex G-150	200,000	150	Cross-linked dextrans (glucose polymers)
Bio-Gel P-100	100,000	100	Cross-linked polyacrylamide
Ultragel AcA34	300,000		Cross-linked polyacrylamide also contains agarose (galactose polymers)

Note: The abbreviation M_r is used to designate the molecular weight of a protein with reference to the mass of a proton.

Problem 4-2

A mixture of protein standards of known molecular weights was eluted from a column containing Sephadex G-200 beads, which have pore size to exclude molecules of approximately 450,000 M_r and above. The presence of protein in a fraction was determined by its absorbance at 280 nm. The elution profile is shown in Fig. 4–3 and the protein standards with their molecular weights are listed in the legend to Fig. 4–3.

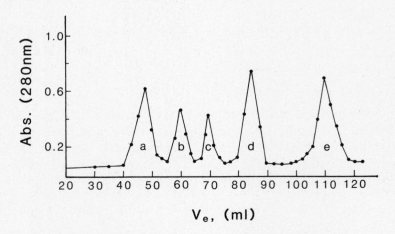

Figure 4-3. Elution profile of protein standards from a Sephadex G-200 column. The proteins are: alcohol dehydrogenase (150,000), bovine serum albumin (66,000), ovalbumin (43,000), carbonic anhydrase (29,000), and myoglobin (17,200). Void volume = 40 mL.

(a) Identify the protein standard corresponding to each peak in the elution profile in Fig. 4–3.

(b) Plot the fraction number (V_e at the peak) for each protein standard as a function of the $\log_{10}$ of its molecular weight. Draw a line of best fit through the points.

(c) Hemoglobin was found to elute from this column with a peak at 60 mL. Using the plot in (b), calculate the approximate molecular weight of hemoglobin.

(d) The enzyme phosphorylase b elutes from this column with a peak at 52 mL. What is its approximate M_r?

Solution:

(a) Since the proteins appear in descending order of M_r, peak a represents alcohol dehydrogenase; b, bovine serum albumin; c, ovalbumin; d, carbonic anhydrase; and e, myoglobin.

(b) See Fig. 4–4.

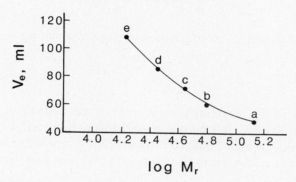

Figure 4–4. Log M_r as a function of elution volume, V_e.

(c) Hemoglobin eluted with a peak at 60 mL, so from Fig. 4–4, the $\log M_r$ must be approximately 4.82. The antilog of 4.82 is 66,000, which is the approximate M_r of hemoglobin. (In the laboratory, the peaks are generally plotted on semilog graph paper so that the molecular weight may be read directly from the graph.)

(d) From Fig. 4–4, the log of the M_r of a protein eluting in fraction 52 (V_e = 52 mL) would be approximately 5.0. The antilog of 5.0 is 100,000, which is the approximate M_r of the phosphorylase b enzyme.

Electrophoresis. Electrophoresis is the process by which molecules are separated on the basis of their net charge. The net charge on a protein is determined by (1) the specific amino acids present and (2) the pH of the solution. The net charge on DNA and RNA fragments is determined almost exclusively by the number of nucleotides present since each PO_4 has a charge of -1 at pH values around 7 and above.

Electrophoretic procedures require that a current of electricity be put through a support medium, which is often a gel formed from polymers of starch or the synthetic compound acrylamide. Polyacrylamide gels are most popular because the cross-linking of the monomers during polymer formation can be more carefully controlled than with starch. The degree of cross-linking determines the porosity, which influences the rate of migration of the molecules being separated.

The net charge on a protein will determine the rate of its migration through the gel as long as the pores in the gel are large enough for the protein to move through easily. The net charge on a protein is related to its isoelectric point (pI) and to the pH of the solution. The pI is defined as that pH at which the molecule has a net charge of zero (see Chapter 2 for a discussion of pI).

In Fig. 4–5, several proteins in a buffer solution of pH 9.5 have been separated on the basis of their net charge. The proteins are detected by staining the gel with Coomassie Blue, a dye that binds to proteins.

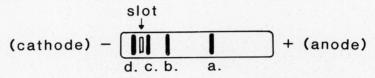

Figure 4-5. Separation of proteins by electrophoresis on polyacrylamide gel.

Problem 4-3

A buffer solution pH 9.5 containing several proteins was added to a slot in a polyacrylamide gel. An electric current (75 V, 0.3 A) was passed through the gel for 4 hr. After staining the gel with Coomassie Blue, which binds to protein, four bands were seen. Given the pI values (below) for the proteins separated in Fig. 4–5, identify each band on the gel diagram.

Protein	pI
Trypsinogen	9.3
Myoglobin	8.0
Carbonic anhydrase	5.0
Cytochrome c	10.7

Solution:

A protein migrates toward the anode (positively charged pole) at a rate proportional to its net negative charge and inversely proportional to its pI value. The protein with the lowest pI value should move the fastest toward the anode (+) since it should have the greatest *net* negative charge ta pH 9.5. Therefore, carbonic anhydrase is represented by band a, myoglobin by band b, and trypsinogen by band c. Band d is a protein that did not move toward the anode but instead migrated a short distance toward the cathode (−), indicating that it had a *net* positive charge at the buffer pH of 9.5. The protein, cytochrome c with a pI of 10.7, would have a small positive charge at pH 9.5 and therefore band d is cytochrome c.

SDS-PAGE (SDS-Polyacrylamide Gel Electrophoresis)

A protein moves in an electric field according to its individual net charge at the pH of the buffer and to its overall size. In order to separate different proteins on the basis of size alone, sodium dodecyl sulfate (SDS) is used to coat the proteins so that each protein's individual charge is not a determining factor in mobility. The SDS molecules contain a hydrophobic tail which interacts with the hydrophobic residues on the protein. SDS also contains two negative charges, and the effect of the SDS coating is to provide negative charges in proportion to the size of the protein. The coating of the protein with SDS destroys the subunit association (quaternary structure) found in many proteins. The SDS-coated proteins are separated by polyacrylamide gel electrophoresis according to size, with the smallest having the greatest R_f value.

$$R_f = \frac{\text{distance traveled by protein}}{\text{distance traveled by dye front}}$$

Different percentages of the acrylamide monomer are used in forming different gels, so that the porosity can be changed to suit the size of the proteins. Large proteins move quickly through a 5% gel but very slowly through an 11% gel.

Problem 4–4

Several proteins of known molecular weight were separated on 7% or 11% SDS-PAGE gels. Diagrams of the gels are shown in Fig. 4–6a (7%) and b (11%).

(a) Identify each protein band by the known M_r of the protein standards.

(b) Plot the log of the M_r of each protein standard as a function of its R_f value.

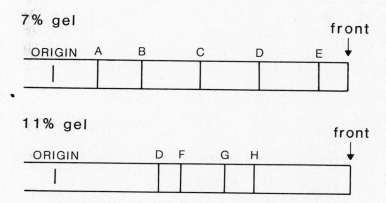

Figure 4–6. (a) Separation of protein standards on 7% gel. Protein standards are: myosin monomer (M_r = 220,000), ovalbumin (M_r = 43,000), β-galactosidase subunit (M_r = 125,000), bovine serum albumin (M_r = 68,000), carbonic anhydrase (M_r = 29,000). (b) Separation of protein standards on 11% gel. Protein standards are: ovalbumin (M_r = 43,000), glycerol-3-PO_4 dehydrogenase (M_r = 36,000), myoglobin (M_r = 17,200), trypsinogen (M_r = 20,100).

Solution:

(a) The proteins move through the SDS gel at a rate inversely proportional to their size. Therefore, the bands in Fig. 4–6 correspond as follows:

(1) 7% gel

Band	Protein	R_f Values	M_r	log M_r
A	Myosin	0.15	220,000	5.34
B	β-Galactosidase subunit	0.30	125,000	5.09
C	Serum albumin	0.50	68,000	4.83
D	Ovalbumin	0.70	43,000	4.63
E	Carbonic anhydrase	0.90	29,000	4.46

(2) 11% gel

Band	Protein	R_f Values	M_r	log M_r
D	Ovalbumin	0.35	43,000	4.63
F	G-3-PDHase	0.42	36,000	4.56
G	Trypsinogen	0.57	20,000	4.30
H	Myoglobin	0.67	17,200	4.23

(b) In Fig. 4–7 the R_f values are plotted as a function of the log of the M_r of each protein. A straight line may be drawn through the data points as a line of best fit. This graph may be used as a standard curve to determine the log of M_r of proteins in the problems that follow.

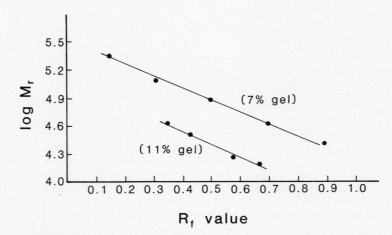

Figure 4-7. R_f values are plotted as a function of log M_r.

Note: Usually, the R_f values for each corresponding M_r (not log M_r) are plotted on semilog graph paper, but the results are the same.

Isolation, Purification, and Analysis of Proteins

During any purification procedure, it is essential that the presence of the desired protein be located after each step. Several different methods are used to detect proteins in solutions and a few are described below.

Detection of Proteins

Two separate properties of proteins are used to detect their presence during purification procedures:

1. The biological function (e.g., an enzyme's ability to catalyze a specific chemical reaction)

2. The chemical properties of a protein (e.g., its reaction with reagents in the Lowry test or its absorbance of light at 280 nm).

Measuring Enzyme Activity. In any cell extract, there are thousands of different proteins but only one or two that catalyze a specific chemical reaction. At each step in the purification process, assays are done to detect the desired enzyme by adding its substrate and measuring the amount of product formed per time unit under defined conditions. The amount of enzyme activity present is often expressed in enzyme units. An *enzyme unit* is defined as

$$1.0 \text{ enzyme unit} = 1.0 \ \mu\text{mol product formed}/1.0 \text{ min}$$

Measuring Protein. The protein in cell extracts and in solutions after purification steps may be measured by using the Lowry test, the Bradford dye binding method, or by reading the absorbance of the solution at 280 nm in a spectrophotometer (see Chapter 1).

Specific Activity. The *specific activity* expresses the amount of enzyme activity as a function of the amount of protein used to catalyze the reaction. Specific activity (S.A.) is defined as

$$\text{S.A.} = \frac{\mu\text{mol product formed/min}}{\text{mg protein}}$$

An increase in the specific activity indicates that the enzyme is "purer" than the original starting enzyme solution. The increase in the S.A. is usually due to a decrease in other proteins which have been removed during the purification process, causing the denominator (of the S.A.) to decrease and thus the S.A. to increase. In a few instances, the increase in the S.A. during the purification of an enzyme is due to the removal of an inhibitor of the enzyme's activity. In this case the increase in the S.A. results from an increase in the amount of product formed per time unit.

Fold purification. The fold purification is determined by dividing the S.A., determined after each purification step, by the S.A. of the beginning solution or cell extract.

Protein Purification Procedures

In a typical protein purification the volume, mg protein, and enzyme activities are measured after each step. The data are presented in a table similar to Table 4-2. In the following experiment, a suspension of red blood cells (2.0 mL of volume) was sonicated to disrupt the cells. The cells were lysed and the resultant cell extract was centrifuged. The supernatant was removed and crystals of ammonium sulfate (AS) were added slowly to make a 0.7 M AS solution. After 30 min, the solution was centrifuged, the supernatant was discarded, and the protein precipitate was dissolved in 1.0 mL of buffer. The S.A. for each step and the fold purification were calculated from the data given in Table 4-2.

Table 4-2. Example of Data from an Enzyme Purification

Purification Step	mg Protein in Assays	μmol P/min	S.A.	Fold Purification
Cell extract	0.10	0.031	$\frac{0.031}{0.10} = 0.31$	0
Supernatant from first centrifugation	0.04	0.023	$\frac{0.023}{0.04} = 0.58$	$\frac{0.58}{0.31} = 1.87$
Dissolved AS precipitate	0.02	0.019	$\frac{0.019}{0.02} = 0.95$	$\frac{0.95}{0.31} = 3.06$

Protein purification procedures are often summarized in flowcharts as shown below. In this flowchart, each step in the procedure described above is indicated and some of the contents of the resulting solutions are listed.

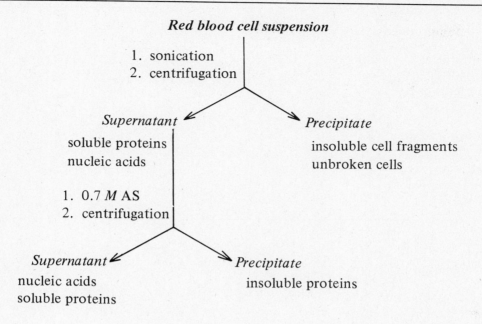

Red blood cell suspension

1. sonication
2. centrifugation

Supernatant

soluble proteins
nucleic acids

Precipitate

insoluble cell fragments
unbroken cells

1. 0.7 *M* AS
2. centrifugation

Supernatant

nucleic acids
soluble proteins

Precipitate

insoluble proteins

Problem 4-5

The isolation of monoamine oxidase from rabbit serum has been described by McEwen[*]. To 1 liter of serum at 4°C is added 200 g of solid ammonium sulfate slowly and with stirring. The resulting solution is kept overnight at 4°C. The precipitate that forms is collected by filtration and discarded. To 500 mL of the filtrate is added 60 g of solid ammonium sulfate. The precipitate contains all the monoamine oxidase activity and is removed by centrifugation at 2000 × g. The resultant precipitate is collected and dissolved in 10 mL of 0.2 *M* ($Na_2HPO_4:NaH_2PO_4$) buffer at pH 7.2. The purification may be summarized as in Table 4-3. The specific activity (S.A.) and the fold purification for the ammonium sulfate purification steps are calculated as follows:

$$S.A. = \mu\text{mol P/min/mg protein} \quad or \quad \frac{\text{enzyme units}}{\text{mg protein}}$$

$$\text{Fold purification} = \frac{\text{S.A. of final solution}}{\text{S.A. of beginning solution}}$$

Table 4-3. Summary of Results in the Purification of Monoamine Oxidase from Rabbit Serum

Step	Total Volume (mL)	Total Enzyme Units	Total Protein (mg)	Specific Activity	Fold Purification
1. Original serum	1000	954	64,000		

[*]C.M.McEwen, "Monoamine Oxidase," in S.P. Colowick and N.O. Kaplan (Eds.), *Methods in Enzymology*, Vol. XVII (B), Academic Press, New York, 1971, pp. 692-698.

Table 4–3 (continued)

2.	Filtrate from first AS addition in 200 mL	200	278	2,083
3.	Filtrate from second AS addition, ppt. dissolved in 10 mL.	12	192	127

Solution:

1. Original serum:

$$S.A. = \frac{954}{64,000} = 0.015$$

2. First ammonium sulfate cut:

$$S.A. = \frac{278}{2,083} = 0.133$$

$$\text{Fold purification} = \frac{0.133}{0.015} = 8.87$$

3. Second ammonium sulfate cut:

$$S.A. = \frac{192}{127} = 1.5$$

$$\text{Fold purification} = \frac{1.5}{0.015} = 100$$

Problem 4-6

Serine dehydratase cleaves the α-amino group from serine, forming NH_3 and pyruvate. This enzyme has been isolated and purified from the bacterium *Clostridium acidiurici* as described by Sagers and Carter.* A suspension of 150 g of *C. acidiurici* cells was made in 300 mL of 0.05 M KH_2PO_4:K_2HPO_4 buffer, pH 7.1. The cells were ruptured by homogenization and the homogenate was centrifuged at 25,000 $\times$ *g* for 20 minutes. Solid ammonium sulfate was added with continuous stirring to give 80% saturation. The precipitate was dissolved in 10 mL of phosphate buffer, and 1.0 mL was added to a Sephadex G-200 column. Fractions of 1.0 mL were collected and assayed for serine dehydratase activity. The activity was found in fractions 26 through 35, with the peak of activity in fractions 30 and 31.

*R.D. Sagers and J.E. Carter, "L-Serine Dehydratase," in S.P. Colowick and N.O. Kaplan (Eds.), *Methods in Enzymology*, Vol. XVII (B), Academic Press, New York, 1971, pp. 351-365.

(a) Prepare a flowchart summarizing this serine dehydratase purification procedure.

(b) From the purification data summarized in Table 4–4, calculate the specific activity and fold purification for each step.

(c) Describe how 1.0 L of the $KH_2PO_4 : K_2HPO_4$ buffer is made. The MW of monobasic KH_2PO_4 is 135.1, and the MW of K_2HPO_4 is 174.2 (see Problem 2–24).

Table 4–4. **Purification of Serine Dehydratase from Clostridium acidiurici**

Step	Volume (mL)	Protein (mg/mL)	Total Enzyme Activity Units	Specific Activity	Fold Purification
Cell-free homogenate	400	11	40,000		
AS ppt dissolved	10	28	11,176		
Sephadex G-200 fraction 30 + 31	2.0	2.4	1,680		

Solution:

(b) Specific activity determinations:

1. Cell-free homogenate:

$$S.A. = \frac{\text{total enzyme units}}{\text{total protein}} = \frac{40,000}{400 \times 11} = \frac{40,000}{4400} = 9.09$$

2. AS precipitate dissolved in buffer:

$$S.A. = \frac{\text{total enzyme units}}{\text{total protein}} = \frac{11,176}{10 \times 28} = 39.9$$

3. Sephadex G-200 fraction 30 + 31:

$$S.A. = \frac{\text{total enzyme units}}{\text{total protein}} = \frac{1680}{2.0 \times 2.4} = 350$$

Fold purification calculations:

1.   Cell-free homogenate: none, by definition

2.   AS precipitate dissolved in buffer:

$$\text{Fold purification} = \frac{\text{S.A. of final solution}}{\text{S.A. of beginning solution}} = \frac{39.9}{9.09} = 4.4$$

3.   Sephadex G-200 fraction 30 + 31:

$$\text{Fold purification} = \frac{\text{S.A. of final solution}}{\text{S.A. of beginning solution}} = \frac{350}{9.09} = 38.5$$

(c) The $KH_2PO_4 : K_2HPO_4$ buffer solution is made in the same way as the $NaH_2PO_4 : Na_2HPO_4$ buffer was made in Problem 2–24.

Isozymes

Isozymes are different proteins that catalyze the same enzyme reaction. In the major metabolic pathways in cells, there are often two or three different enzymes that will catalyze each reaction. In many instances, isozymes are different proteins with very similar amino acid sequences coded for by different, but closely related genes.

The detection of isozymes usually follows acrylamide gel electrophoresis, where the gel is stained to detect enzyme activity as shown in Fig. 4–8a. The presence of more than one band indicates different proteins with the same enzyme activity (i.e., isozymes).

The enzyme lactate dehydrogenase (LDH) is found in most cells and catalyzes the reaction

$$\text{pyruvate} + \text{NADH} + H^+ \xrightarrow{\text{LDH}} \text{lactate} + NAD^+$$

The presence of LDH isozymes has been demonstrated by the multiple bands of LDH activity found after electrophoresis of liver extracts. In Fig. 4–8 identical gels were stained to detect (a) the presence of LDH activity and (b) total protein. As one would expect, there are hunderds of proteins in a partially purified rat liver extract (Fig. 4–8b), but in gel a, only those proteins that are able to catalyze the formation of lactate and NAD^+ from pyruvate and NADH are visible. Five different proteins with LDH activity are obvious in Fig. 4–8a, indicating that there are five isozymes for LDH in rat liver extracts.

Estimating the Size of a Protein

The molecular weight (M_r) of a pure protein is often estimated by gel exclusion chromatography using beads with pores large enough for the protein to penetrate. Beads such as Sephadex G-200, Sepharose 6, and Ultragel A2A34 have pores that many large proteins may enter. By measuring the elution volume (V_e) for proteins of known molecular weight (Fig. 4–3), a standard curve may be developed as in Fig. 4–4, where the log of the M_r of a protein is plotted as a function of the V_e for its peak fraction. Then a solution of a pure protein of unknown M_r may be run *on the same column* and the V_e of its peak fraction used to find the log of its M_r. Other methods used to determine the M_r of a pure protein are ultracentrifugation and SDS-polyacrylamide gel electrophoresis, but gel exclusion chromatography is perhaps the method most frequently used.

Figure 4-8. Polyacrylamide gels showing separation of proteins from the same cell extract: gel a is stained to detect LDH activity only; gel b.shows all proteins.

Problem 4-7

In experiments using aliquots of 0.5 mL of rat liver extract in gel exclusion chromatography, it has been shown that only one peak of lactate dehydrogenase activity elutes from the Sephadex G-200 column, while hundreds of other proteins are present in the elution profile as shown in Fig. 4–9. The peak of the LDH enzyme activity is seen in fraction, V_e = 50 mL. From the elution profile for LDH activity in Fig. 4–9 and the data in Fig. 4–8a, what can you conclude about the LDH isozymes?

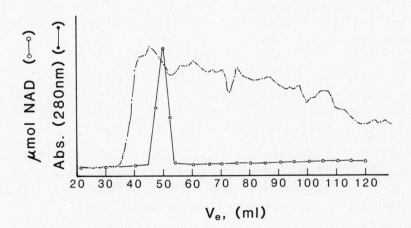

Figure 4-9. Elution profile of LDH activity (o——o) and total protein (•——•) from Sephadex G-200 column.

Solution:

Since there are five distinct bands of LDH activity after acrylamide gel electrophoresis, there must be five different proteins (isozymes) which are capable of catalyzing the LDH reaction. The elution profile in Fig. 4–9 indicates that each of these five isozymes has the same M_r or very close to the same M_r. From Fig. 4–4, a protein eluting with V_e = 50 mL has log M_r = 5.0. The antilog of 5.0 = 100,000. Therefore, the approximate M_r of each of the 5 LDH isozymes is 100,000.

Determining the Subunit Structure of a Protein

Many proteins are multimeric; that is, they are composed of several subunits, which may or may not be identical polypeptides. The enzyme hexokinase consists of two identical subunits, but pyruvate dehydrogenase found in beef heart mitochondria is composed of 160 subunits of six different types.

The M_r of the native multimeric protein may be found by gel exclusion chromatography. The different types of subunits and their molecular weights are determined by SDS-polyacrylamide gel electrophoresis (SDS-PAGE). Proteins that are single polypeptides give the same M_r values by gel exclusion chromatography and SDS-PAGE. However, multimeric proteins are dissociated by SDS treatment, giving different M_r values corresponding to the M_r of the subunits. The number of different bands on SDS-PAGE usually equals the number of different subunits present in the multimeric protein.

Problem 4–8

Aldolase is an enzyme that catalyzes the cleavage of F-1,6-bisP in glycolysis. Aldolase is composed of four identical subunits each of M_r = 40,000.

(a) In which fraction would the native aldolase elute from the Sephadex G-200 column in Problem 4–2 (Fig. 4–3)?

(b) If pure aldolase were used in the separation by SDS-electrophoresis (7% gel) in Problem 4–4 (Fig. 4–6a), how many bands would appear, and what would the R_f value(s) be?

Solution:

(a) The native form of aldolase is a tetramer of four identical subunits, M_r = 40,000 and therefore native aldolase has M_r = 160,000. The log of 160,000 = 5.20. From Fig. 4–4, proteins of M_r = 160,000 should elute with a peak at about 45 mL.

(b) Since pure aldolase is a tetramer of four identical subunits, only one band should be seen in SDS gels. The R_f of that band is obtained from Fig. 4–7, in which the log of the M_r is plotted as a function of the R_f value. The log of 40,000 = 4.60 and extrapolating from the graph, the R_f value would be approximately 0.77.

Problem 4–9

Pure hemoglobin has an R_f value of 0.67 in an 11% gel in an SDS-polyacrylamide gel electrophoresis experiment, similar to the one described in Problem 4–4.

(a) What is the M_r of hemoglobin from this R_f value on SDS-gel electrophoresis?

(b) Does this agree with the M_r obtained for hemoglobin by gel exclusion chromatography in Problem 4–2? Explain the discrepancy.

Solution:

(a) From Fig. 4–7, proteins with an R_f value of 0.67 would have a log M_r = 4.22. The antilog of 4.22 = 16,595, so that the approximate M_r of hemoglobin by SDS-gel analysis is 16,600.

(b) The M_r of hemoglobin determined by gel exclusion chromatography (Problem 4–2) was 66,000. The SDS-gel data suggest that native hemoglobin is composed of four subunits of similar M_r. The subunits may or may not be identical from the SDS-gel data. If the subunits were identical only one band would be found on SDS-gel electrophoresis as was found here. But if the subunits were different proteins with very similar M_r, only one band would appear on SDS-gel electrophoresis.

Problem 4–10

The native enzyme β-galactosidase elutes from the Sephadex G-200 column in Fig. 4–3 with a peak at 39 or 40 mL. What can you conclude about the structure of β-galactosidase from its R_f value on SDS-gel electrophoresis (Figs. 4–6a and 4–7) and its elution from the Sephadex G-200 column?

Solution:

Since the bed volume (volume of the bed of Sephadex G-200 beads) is 40 mL, it would appear that the native β-galactosidase enzyme is too large to enter the beads and thus probably has a M_r greater than 450,000, but how much greater cannot be determined from these data. The R_f value from SDS-gel electrophoresis indicates the M_r to be 125,000 (Problem 4–4). Together, these experiments show that native β-galactosidase is probably composed of at least four subunits of M_r = 125,000. Actually, from other experiments, it was determined that β-galactosidase from *Escherichia coli* bacteria has M_r = 500,000 and is a tetramer of four identical subunits, each with M_r = 125,000.

Problem 4–11

The enzyme creatine kinase catalyzes the formation of phosphocreatine from creatine and ATP. Phosphocreatine supplies the energy required for contraction of mammalian muscle. Creatine kinase was isolated and purified from an extract of beef heart. Pure creatine kinase from beef heart extracts were electrophoresed on polyacrylamide gels, and three separate bands with creatine kinase activity were always demonstrated. Analysis of fractions from the Sephadex G-200 column in Fig. 4–3, loaded with creatine kinase, showed one broad peak of enzyme activity (peak at V_e = 53 mL). SDS-gel electrophoresis of a pure preparation of creatine kinase revealed two discrete bands on an 11% gel (Fig. 4–6b), one band with an R_f value of 0.35 and the other, 0.32.

What can you conclude from these data about the structure of creatine kinase?

Solution:

The data from gel exclusion chromatography using Sephadex G-200 indicate that the native enzyme has an M_r = the antilog of 4.95, which is about 90,000. Since the peak is broad, there may be more than one form of the enzyme and each with slightly different M_r. The three bands of creatine kinase activity obtained after polyacrylamide gel electrophoresis strongly suggests that this enzyme exists in beef heart as three isozymes. The two discrete bands obtained on SDS gels show that the isozymes are composed of two different subunits with M_rs equal to the antilogs of 4.63 and 4.65, that is, M_r = 43,000 (subunit A) and 46,000 (subunit B), respectively.

Creatine kinase appears to exist in three forms (isozymes), each with a M_r *close* to 90,000 and each consisting of two subunits. Since the two subunits are different proteins, the most probable structure of the three different isozymes is: AA, AB, and BB. In this case, the M_r of the three isozymes would be: AA, M_r = 86,000; AB, M_r = 89,000; and BB, M_r = 92,000. These three isozymes would not give a clear separation on a Sephadex G-200 column but rather would give a broad peak around V_e = 53 mL.

Determining Whether Subunits are Identical

Some multimeric proteins are composed of identical subunits, whereas others are composed of different polypeptide subunits. If only one band is found for a protein in SDS gels, it may be concluded that the subunits have the same M_r. However, several instances are known where different subunits had identical M_r's. An indication of whether the subunits in a protein are identical may be obtained by determining the N-terminal sequence of each polypeptide using the Edman procedure. The Edman reagent reacts with the free amino group at the N terminal of each polypeptide. The Edman derivative of the N-terminal amino acid may be removed and identified, exposing a new N-terminal amino acid which then may be identified by this Edman method. Only short regions of polypeptides may be sequenced at one time in this way, but in just a few residues at the N terminal it may be obvious that the subunits are not identical, as in Problem 4–12.

A major difficulty in using N-terminal analysis to determine subunit similarities is due to the normal alteration of the N terminal within the cell by proteases, acetylases, and other enzymes, which modify the N-terminal end. As many as 50% of the proteins in a eukaryotic cell have acetyl groups or other small molecules covalently bound to the N-terminal amino group, preventing it from reacting with reagents such as the Edman reagent.

Problem 4–12

Enzymes called luciferases catalyze the production of light in bioluminescent organisms such as fireflies and luminescent bacteria. The reaction catalyzed by firefly luciferase enzyme requires ATP and luciferin, a complex heterocyclic compound which is oxidized and in the process, gives off a flash of light. The reaction in luminescent bacteria does not require ATP. The light energy given off as a glow by these bacteria is obtained from the oxidation of a long-chain saturated aldehyde and reduced $FMN \cdot H_2$ to yield the corresponding carboxylic acid, blue-green light (maximum = 490 nm), and FMN.

Bacterial luciferase was purified from the luminescent bacteria *Photobacterium fischeri*, and the results of several experiments are described below. [Data from T. O. Baldwin, et al., 1979, *Proc. Natl. Acad. Sci. (USA)* 76: 4887–4889.]

1. This bacterial luciferase eluted from the Sephadex G-200 column in Fig. 4–3 with a peak at 55 mL.

2. On 11% SDS gels, pure luciferase formed two bands with R_f values of 0.37 and 0.40.

3. Edman degradation of the pure enzyme yielded two slightly different N-terminal sequences:

    (a) N-met-lys-phe-gly-asn-ile-ser-phe-ser-tyr-

    (b) N-met-lys-phe-gly-leu-phe-phe-leu-asn-phe-

What can you conclude about the structure of this luciferase from these data?

Solution:

The native protein elutes from the Sephadex G-200 column with a peak at 55 mL. From Fig. 4–4 this corresponds to a M_r equal to the antilog of 4.90, which is approximately 79,400. The bands on the 11% SDS gel at R_f values 0.37 and 0.40 indicate that there are two different subunits with M_r's of 41,000 and 38,000. The amino acid sequence data for the two subunits confirms that the subunits are different but their amino acid sequences are quite similar, at least at the N terminal.

Determining the Amino Acid Sequence of a Polypeptide

To determine the sequence of the amino acids in proteins, several milligrams of pure protein are usually required for each of the following steps.

Note: All the disulfide bridges between cys residues must be broken *before* the next steps can be performed. Disulfide bonds may be broken by treating the protein with mercapto-ethanol.

Step 1: Determining the Amino Acid Composition of the Protein. All the peptide bonds must be broken by heating the protein in strong acid. The free amino acids that are released may be separated and quantitated using an amino acid analyzer. The results show which amino acids are present and in what quantities, but do not give information as to their sequence in the protein.

Step 2: Identifying the Amino and Carboxyl Terminals. Specific reagents may be used to cleave the amino acid residues from either the N- or C-terminal ends. The cleaved residues may then be identified by different methods of chromatography.

Identifying the N-terminal Amino Acid.   *Dansyl chloride* reacts with free N-terminal residues, and the dansyl derivative of the N-terminal amino acid may be identified by chromatography after hydrolysis of all peptide bonds.   The *Edman reagent* labels and removes the free N-terminal residue and allows identification of the amino terminal without destroying the rest of the protein. This is the preferred method of N-terminal identification. In eucaryotic cells, many of the proteins have acetyl groups covalently bound to the N-terminal amino group, preventing reaction with dansyl chloride or the Edman reagent. The acetyl group or the N-terminal amino acid (with its acetyl group) must be removed before sequencing can proceed from the N-terminal end.

Identifying the C-terminal Amino Acid. The C-terminal amino acid may be cleaved from the protein by one of the different forms of the enzyme carboxypeptidase. Carboxypeptidase A is obtained from beef pancreas and cleaves the C-terminal amino acid except when it is arg, lys, or pro. Carboxypeptidase B cleaves only arg or lys N terminals from proteins and is purified from pork pancreas. Neither enzyme cleaves C-terminal amino acids when pro is the adjacent residue.

Step 3: Fragmenting the Polypeptide and Isolating the Fragments. Several cleavage procedures may be used to break the polypeptide into smaller fragments, which may be sequenced more easily. Enzymes (proteases) that cleave peptide bonds between specific amino acids are used to treat the polypeptide to give peptide fragments. The enzymes most commonly used are shown in Table 4–5 with their specificities. Proteins are treated with the reagent cyanogen bromide (CNBr), which cleaves peptide bonds in which the carbonyl is from methionine.

Table 4–5. Enzymes Most Commonly Used to Cleave Proteins

Enzyme	Specificity[a]	Peptide bond cleaved
Trypsin	arg or lys residues	Carboxyl of arg or lys (except when pro is adjacent)
Chymotrypsin	phe, tyr, trp, leu, ile, val	Carboxyl of these residues (except when pro is adjacent)
Thermolysin	met, ile, leu, or any hydrophobic residue	N of met, ile, leu or any hydrophobic residue

[a] Indicates residues most commonly found at cleavage sites.

Note: In many general textbooks, the specificity of chymotrypsin is limited to the aromatic residues, for the sake of simplicity.

The fragments produced by any of the digestive processes described above may be separated using chromatographic techniques. Each fragment may be treated to digest it further.

Step 4: Sequencing the N-terminal Portion of a Fragment Using the Edman Degradation Technique. Under controlled conditions the Edman degradation procedure may be used to label and remove up to 15 consecutive amino acids at the N-terminal end of the protein. Under optimal conditions, the Edman degradation scheme may be used to determine the N-terminal sequence of a polypeptide up to about 20 residues.

Step 5: Ordering the Peptide Fragments by Establishing Overlaps and Finally, Determining the Sequence of Amino Acids Using All the Data. The following polypeptide was sequenced and the results are shown below.

$$NH_3\text{-met-gly-ala-trp-lys-asp-his-met-phe-arg-glu-tyr-cys-ser-leu-COO}^-$$

Data obtained from the steps described above.
From step 1, amino acid composition:

trp:ser:tyr:phe:ala:arg:asp:cys:gly:glu:his:leu:lys:met

  1   1   1   1   1   1   1   1   1   1   1   1   1   2

From step 2:

1. The Edman reagent cleaved met from the N terminus of the polypeptide.

2. Carboxypeptidase A cleaved leu from the C terminus.

From step 3: Incubation of an aliquot of the polypeptide with trypsin yielded three peptide fragments, incubation with chymotrypsin gave four peptide fragments, and treatment with cyanogen bromide produced two peptide fragments.

From step 4: Edman degradation of these peptides gave the following amino acid sequences:

Tryptic digests:	fragment	1 glu-tyr-cys-ser-leu
		2 met-gly-ala-trp-lys
		3 asp-his-met-phe-arg
Chymotryptic digests:	fragment	1 cys-ser-leu
		2 met-gly-ala-trp
		3 lys-asp-his-met-phe
		4 arg-glu-tyr

From step 5, ordering the peptide fragments and analyzing all the data:

N terminus   C terminus

met   leu

Tryptic digests:

met-gly-ala-trp-lys   glu-tyr-cys-ser-leu

Chymotryptic digests:

lys-asp-his-met-phe   arg-glu-tyr

Problem 4-13

Venom has been collected from thousands of honeybees and analysis has shown it to contain the protein melittin, in addition to the amino acid derivatives histamine and dopamine, and the enzymes phospholipase and hyaluronidase. The protein melittin causes the intense pain a person feels immediately after being stung by a honeybee, but the swelling and redness that appear later are due largely to histamine, not to melittin. The active form of melittin appears to be a tetramer of identical subunits which stimulate certain nerve cells, causing intense pain. The amino acid sequence of a melittin subunit has been determined (Lubke et al., 1971, *Experientia 27*:765-767) and is as follows:

N-gly-ile-gly-ala-val-leu-lys-val-leu-thr-thr-gly-leu-pro-ala-leu-ile-ser-trp-ile-lys-arg-lys-arg-gly-gln-COO$^-$

Show what fragments would be obtained if purified melittin were digested with:

(a) Trypsin

(b) Chymotrypsin

(c) Cyanogen bromide (CNBr)

(d) Carboxypeptidase A

Solution:

(a) Tryptic fragments:

gly-ile-gly-ala-val-leu-lys

val-leu-thr-gly-leu-pro-ala-leu-ile-ser-trp-ile-lys

gln-gln

In addition to these peptide fragments, free arg and free lys are also found in tryptic digests of melittin.

(b) Chymotryptic fragments (in addition to free leu and ile:)

gly-ile, gly-ala-val, lys-val, thr-thr-gly-leu

pro-ala-leu, ser-trp

lys-arg-lys-arg-gln-gln

(c) Cyanogen bromide (CNBr) does not cleave this polypeptide since CNBr acts on internal met residues.

(d) Carboxypeptidase A cleaves gln from the C-terminal end of melittin.

Problem 4-14

A small peptide has been isolated which has the same properties as an endorphin. After acid hydrolysis, the amino acid composition was found to be

1 arg: 2 gly: 1 leu: 2 lys: 1 phe: 1 pro: 2 tyr

The dansyl chloride derivative was found to be tyr. Neither carboxypeptidase A nor B cleaved the peptide. After digestion of the peptide by trypsin, one large peptide fragment, free lys, and a tripeptide were isolated. Using the Edman degradation method, the tripeptide was shown to be tyr-pro-lys, and the larger peptide, tyr-gly-gly-phe-leu-arg. Chymotrypsin digestion yielded two peptide fragments and free leu and tyr. One fragment from the chymotryptic digest was treated with Edman reagent and had the following sequence: arg-lys-tyr-pro-lys.

(a) From the data above, determine the amino acid sequence of this peptide.

(b) Why do neither of the carboxypeptidases cleave this peptide?

Solution:

(a) The amino acid sequence was deduced from the data as shown below.

N terminal:

tyr

Tryptic peptides:

tyr-pro-lys   tyr-gly-gly-phe-leu-arg

It is not clear from the tryptic digests which of these fragments is the N-terminal peptide.

Chymotryptic peptides:

arg-lys-tyr-pro-lys

This chymotryptic peptide indicates that the tryptic peptide tyr-pro-lys must *not* be the N-terminal peptide and, in fact, is probably the C terminal. The free lys in the tryptic digests probably lies between arg (C terminal of the large fragment) and tyr (N terminal of the tripeptide).

Thus the amino acid sequence for this polypeptide is most likely

N-tyr-gly-gly-phe-leu-arg-lys-tyr-pro-lys-COO⁻

(b) Carboxypeptidases will not cleave the C terminal when pro is the adjacent amino acid.

Problem 4–15

The small protein dynorphin B isolated from the bloodstream of pigs appears to have an opiate-like effect on the human brain. When completely hydrolyzed, the amino acid composition was determined to be

arg:gly:gln:lys:leu:phe:thr:tyr:val

2   2   1   1   1   2   1   1   2

The dansyl chloride derivative was that of tyr and carboxypeptidase A cleaved thr from the pure protein. Trypsin digestion yielded free arg, two small peptides, and one large peptide. Edman degradation of the small peptides revealed these sequences:

gln-phe-lys   tyr-leu-arg

Chymotrypsin digestion yielded free tyr, leu, val, and thr, and two peptide fragments. Edman degradation showed the peptides to have the following sequences:

lys-gly-gly-phe   arg-arg-gln-phe

Deduce the sequence of amino acids in this protein.

Solution:

N terminal   C terminal

tyr   thr

Tryptic peptides:

tyr-leu-arg   gln-phe-lys

Chymotryptic peptides:

arg-arg-gln-phe   lys-gly-gly-phe

Free tyr must be the N terminal; the free thr is the C terminal; the free val is between phe and thr. Then the sequence must be

NH_3-tyr-leu-arg-arg-gln-phe-lys-gly-gly-phe-val-val-thr-COO^-

Problem 4–16

The small peptide γ-melanocyte-stimulating hormone (MSH) is known to cause darkening of the pigmented areas of the skin in most vertebrates. MSH has been purified and analyzed as shown below. The amino acid composition was shown to be

2 arg: 1 asp: 2 gly: 1 his: 1 met: 2 phe: 1 tyr: 1 trp: 1 val

Edman reagent derivative = tyr. Carboxypeptidase A cleaved gly. Trypsin digests of MSH produced three different peptide fragments, two of which were shown to have the sequences

trp-asp-arg   phe-gly

Chymotrypsin digests yielded free gly, free tyr, free val, a dipeptide (arg-trp), a tripeptide, and a peptide fragment (met-gly-his-phe). From these data, determine the sequence of amino acids in MSH.

Solution:

N terminal   C terminal

tyr   gly

Trypsin fragments:

(large fragment)   trp-asp-arg   phe-gly

Chymotrypsin fragments:

met-gly-his-phe   arg-trp

Then in the chymotrypsin digests the free gly must be the C terminal, the free tyr is the N terminal, and val must be between tyr and met. Then the amino acid sequence of MSH is

NH_3^+-tyr-val-met-gly-his-phe-arg-trp-asp-arg-phe-gly-COO^-

Problem 4–17

Bradykinin is a small polypeptide found in mammalian blood. It causes dilation of blood vessels, especially in the kidney. The following data were obtained by analysis of pure bradykinin:

The amino acid composition of bradykinin was found to be

2 arg: 1 gly: 3 pro: 2 phe: 1 ser

Dansyl chloride derivative was arg. Carboxypeptidase B cleaved arg from bradykinin. Trypsin did not cleave this peptide. Chymotrypsin digests yielded free arg and two peptide fragments, which when sequenced with Edman procedure, gave

ser-pro-phe   arg-pro-pro-gly-phe

(a) From these data, determine the amino acid sequence of bradykinin.

(b) Why did trypsin not cleave this peptide containing two arg residues?

Solution:

N terminal   C terminal

arg   arg

Since trypsin did not cleave N-terminal arg, this suggests the residue next to the N terminal may be pro. Chymotrypsin fragments:

arg-pro-pro-gly-phe   ser-pro-phe   arg

(a) The sequence is:

N-arg-pro-pro-gly-phe-ser-pro-phe-arg-C

(b) Since pro follows the N-terminal arg, trypsin cannot cleave the peptide bond.

Problem 4–18

The polypeptide somatostatin regulates growth in humans and is termed the human growth hormone. Pure somatostatin has been hydrolyzed and found to be composed of

1 ala: 1 asn: 2 cys: 1 gly: 2 lys: 3 phe: 1 tyr: 1 trp: 1 ser: 1 thr

Digestion of somatostatin by chymotrypsin gave free phe, free trp, free tyr, two small peptide fragments, and one large peptide. Edman degradation of the two small peptide fragments gave the following sequences:

lys-thr-phe   ser-cys

Digests of the large peptide fragment (from chymotryptic digest) with trypsin yielded two smaller peptides, which Edman degradation revealed to have the following sequences:

asn-phe   ala-gly-cys-lys

Trypsin digests of whole somatostatin peptide gave three peptides, one of which had the amino acid sequence

thr-phe-tyr-ser-cys

The dansyl chloride derivative of ala was obtained, and carboxypeptidase A treatment produced free cys. From these data, deduce the amino acid sequence of human somatostatin.

Solution:

N terminal   C terminal

ala   cys

Chymotryptic digests:

lys-thr-phe   ser-cys

Tryptic digest of large chymotryptic fragment of somatostatin:

ala-gly-cys-lys   thr-phe-tyr-ser-cys

The fragment asn-phe must lie between lys and lys together with another phe and trp. The sequence is

ala-gly-cys-lys-asn-phe-phe-trp-lys-thr-phe-tyr-ser-cys

or, ala-gly-cys-lys-asn-phe-trp-phe-lys-thr-phe-tyr-ser-cys

Problem 4–19

Certain normal genes in the human genome may mutate to give a mutant form which is oncogenic; that is, it may cause the mutant cell to become cancerous. In human cells, the normal gene, ras p21, codes for a protein of 189 amino acid residues which apparently acts as part of a growth-promoting complex in cells. Some mutant forms of this gene alter the protein so that it loses its ability to be regulated and causes that cell to escape normal control of cell division, thus becoming cancerous. The ras p21 protein, isolated from a normal human cell, has been purified and trypsin digests gave several fragments (B. Jackson, et al., 1985, *Trends Biochem. Sci. 10*:350–353). Normal ras p21 protein N-terminal sequence:

$$1$$

Fragment N_1: NH_3^+-met-thr-glu-tyr-lys-

$$6 \qquad\qquad 10 \qquad\qquad\qquad 16$$

Fragment N_2:   leu-val-val-val-gly-ala-gly-gly-val-gly-lys-

A mutant ras p21 protein purified from a human bladder carcinoma also had 189 residues. The fragment N_2 was absent in tryptic digests of this mutant form, but two new fragments were found, which had the following sequences:

leu-val-val-val-gly-ala-arg   gly-val-gly-lys

From these amino acid sequence data, determine what single base change in the DNA of the ras p21 gene gave rise to this oncogenic ras p21 form, (see Figure 4–10).

Solution:

The new tryptic digest fragments in the mutant ras p21 are similar to the normal N_2 fragment, except that arg is substituted for gly at position 12. The substitution of an arg for a gly would give a new site for trypsin cleavage and thus produce two fragments in place of the N_2 fragment. The single base change in the gene for ras p21 may be determined by comparing the codons for arg and gly.

Codons for gly: GGA, GGG, GGC, and GGU

Codons for arg: AGA and AGG; CGA, CGG, CGU, and CGC

The single base change that gave rise to this oncogenic ras p21 must have been the substitution of T or G for C in the first position for residue 12.

Normal $\longrightarrow$ Mutant

protein: gly $\longrightarrow$ arg

mRNA: GGA $\longrightarrow$ AGA   or GGA $\longrightarrow$ CGA

DNA: CCT $\longrightarrow$ TCT   or CCT $\longrightarrow$ GCT
(sense strand)

Purification and Analysis of DNA

DNA generally occurs as a double-stranded helix where the two strands are held together by H bonds and hydrophobic interactions between the purine and pyrimidine bases. There is a negative charge on each phosphate group of the phosphoester backbone, and in the eukaryotic cell, histone proteins (rich in arg and lys) are attracted to these PO_4^- charges, coating portions of the DNA and forming the complex known as a nucleosome. Eukaryotic chromosomes are composed of nucleosomes and linker DNA. RNA is almost always single-stranded but exists in nature as a twisted strand with base pairing occurring between different regions of the same strand.

DNA Purification

In purifying the DNA from disrupted cells, the protein and RNA must be removed from the DNA. Protein is often precipitated by "salting out" in high salt solutions while the DNA remains quite soluble. DNA may be precipitated from solution by adding cold absolute ethanol slowly. Generally, DNA precipitated in this way is relatively free of RNA. The DNA is dissolved in buffer at pH 7.5 and checked for purity. If protein remains as a contaminant, it may be removed by treatment with phenol or by incubating the preparation with protease enzymes which degrade proteins but not nucleic acids. RNA is removed by treating the extract with purified ribonuclease, an enzyme that degrades RNA but not DNA.

In some cases, isolating, purifying, and sequencing DNA for a specific gene may be accomplished in a few months, whereas a few years are often required to purify and sequence a protein. Since the sequence of bases in the DNA codes for the sequence of amino acids in proteins, it is becoming increasingly easier to determine the base sequence in the DNA and, using the codon dictionary (Fig. 4–10), decipher the amino acid sequence of the protein in this way. In eucaryotes, cDNA made from a mRNA must be sequenced rather than the actual DNA, so that only the exons are used.

Problem 4–20

Determine the amino acid sequence in the protein that is coded for by the following DNA sequence, using the codon dictionary (Fig. 4–10). The DNA sense strand is

$3'$-OH   TACATGCTAATGCGCTATTCTAAT $5'$-p

Solution:

The mRNA would be transcribed beginning at its $5'$-p end, which is complementary to the $3'$-OH end of DNA.

mRNA: $5'$-p   AUGUACGAUUACGCGAUAAGAUUA $3'$-OH

The protein coded for by this mRNA is determined by translating the codons into amino acids, beginning at the $5'$-p end of the message.

peptide: NH_3^+ met-tyr-asp-tyr-ala-ile-arg-leu COO^-

Figure 4–10. The Genetic Code

5'-P Terminal Base	Middle Base				3'-OH Terminal Base
	U	C	A	G	
U	Phe	Ser	Tyr	Cys	U
	Phe	Ser	Tyr	Cys	C
	Leu	Ser	Term[b]	Term[b]	A
	Leu	Ser	Term[b]	Trp	G
C	Leu	Pro	His	Arg	U
	Leu	Pro	His	Arg	C
	Leu	Pro	Gln	Arg	A
	Leu	Pro	Gln	Arg	G
A	Ile	Thr	Asn	Ser	U
	Ile	Thr	Asn	Ser	C
	Ile	Thr	Lys	Arg	A
	Met[a]	Thr	Lys	Arg	G
G	Val	Ala	Asp	Gly	U
	Val	Ala	Asp	Gly	C
	Val	Ala	Glu	Gly	A
	Val[a]	Ala	Glu	Gly	G

[a]Sometimes used as initiator codons. When GUG is used as an initiation codon, it codes for Met.

[b]Stands for termination or stop codon.

Amino acid abbreviations: Phe (phenylalanine); Leu (leucine); Ile (isoleucine); Met (methionine); Val (valine); Ser (serine); Pro (proline); Thr (threonine); Ala (alanine); Tyr (tyrosine); His (histidine); Gln (glutamine); Asn (asparagine); Lys (lysine); Asp (aspartic acid); Glu (glutamic acid); Cys (cysteine); Trp (tryptophan); Arg (arginine); Gly (glycine).

Problem 4–21

Human growth hormone (somatostatin) is required to treat growth deficiencies in several hundred children each year. Since only a few nanograms are obtained from each deceased human donor and microgram quantities are required to treat a growth-hormone-deficient child, this polypeptide has been the object of cloning efforts for several years. Since the amino acid sequence was determined and it is such a small peptide, the cDNA for somatostatin was synthesized using a base sequence determined from the known amino acid sequence given below.

Using the codon dictionary (Fig. 4–10), determine one DNA sequence that will code for this polypeptide.

Note: There is more than one sequence that will code for somastatin.

Amino acid sequence for somatostatin:

N-ala-gly-cys-lys-asn-phe-phe-trp-lys-thr-phe-tyr-ser-cys-C

Solution:

The antisense DNA strand will have the same base sequence as the mRNA except that the U's will be T's.

mRNA strand:

5'-GCUGGUUGUAAAAACUUUUUCUGGAAGACAUUUUAUAGUUGC-3'

Antisense DNA strand:

5'-GCTGGTTGTAAAAACTTTTTCTGGAAGACATTTTATAGTTGC

Sense DNA strand:

3'-CGACCAACATTTTTGAAAAGACCTTCTGTAAAATATCAACG-5'

Problem 4–22

A biochemist wishes to synthesize a piece of the DNA in the promoter region of the gene coding for an enzyme that detoxifies hydrocarbons in the human lung. The desired sequence is that for the sense strand of the DNA. The sequence for the antisense strand has been found in a recent paper and is shown below. Give the sequence of the sense strand starting at the 5' end.

Antisense DNA strand:

5'-TCGTTTATCTTAAATTCCAC-3'

Solution:

The pairing of the strands of the DNA double helix are antiparallel so that the 5' end of the antisense strand pairs with the 3' end of the sense strand. Therefore, the desired sequence is:

3'-AGCAAATAGAATTTAAGGTG-5'

Detection of DNA

Absorbance in the Ultraviolet Range. The bases in DNA absorb light in the ultraviolet (UV) range between 225 and 290 nm and therefore allow DNA in solution to be detected using a spectrophotometer. Unfortunately, RNA also absorbs light at these same wavelengths and any RNA contaminants will add to the absorbance.

Binding of Ethidium. Ethidium binds to nucleic acids by intercalating between the bases. Under UV light, ethidium fluoresces a bright pink color and is used to locate both RNA and DNA on agarose gels.

Characterization of Purified DNA

Total DNA isolated from cells is termed *genomic DNA* and consists of millions of fibers of more than 10,000 base pairs (10 kilobase pairs, or 10 kb) in length. These high-molecular-weight DNA molecules may be cleaved to produce fragments that may be used to determine the base sequence.

Fragmenting the DNA. Specific enzymes, termed *restriction endonucleases,* are used to cleave the phosphoester bonds at specific base sequences (Table 4–6). These cleavages produce fragments of DNA that may be separated easily using electrophoresis in an agarose gel.

Table 4–6. Restriction Endonucleases Commonly Used to Cleave DNA

Enzyme Designation	Base Sequence Specificity (5' to 3')[a]
Alu I	AG/CT
Bam H1	G/GATCC
Eco R1	G/AATTC
Hind III	A/AGCTT
Hpa I	GTT/AAC
Hpa II	C/CGG (will not cleave CC*GG)
Msp I	C/CGG (cleaves CC*GG)
Pst I	CTGCA/G

[a] An asterisk indicates 5-methyl cytosine.

Determining the Size of DNA Fragments. The fragments produced by digesting the DNA with specific restriction enzymes may be separated on the basis of their size by agarose gel electrophoresis. In these gels, the distance traveled by the DNA fragment in the electric field is inversely proportional to its size. Thus the size of a DNA fragment in terms of the number of nucleotides is determined.

Note: Only the number of nucleotides present in each fragment may be determined from the gels. The nucleotides contained in the DNA fragment are determined only by deducing the sequence based on the treatment that produced the fragment.

Restriction Site Maps of DNA. Restriction endonucleases bind to specific base sequences in DNA molecules and cleave the phosphoester bonds in much the same way as proteases bind specific amino acid sequences in proteins and cleave the peptide bonds. The specific base sequences may be "mapped" along the DNA molecule by comparing the size of the DNA fragments formed after digestion with a restriction endonuclease for which the base specificity is known.

1. The first step in restriction site mapping is to set the reference point for the restriction sites. This is done by digesting the intact circular DNA molecules with a restriction enzyme that has only one binding site and therefore cleaves in only one position. When both strands of the circular DNA are cleaved in one position, a linear DNA molecule is formed. Electrophoresis of the digested DNA will show only one band.

2. The relative positions of other restriction sites are determined by combining two or more different restriction enzymes in the same digestion and comparing the sizes of the fragments after separation by electrophoresis. It is important to understand that the base sequence of each fragment is not determined from this technique, only the size in base pairs.

As an example, we can analyze the results of digestions of the plasmid YIP5 by several restriction endonucleases, each with a different base sequence specificity. YIP5 exists as double-stranded DNA of 5541 base pairs in a circle and is used as a cloning vehicle to transfer genes into yeast cells.

YIP5 has only one recognition site for the restriction enzyme Eco R1, which cleaves at the DNA sequence GAATTC. Then we arbitrarily set the first G in the Eco R1 recognition sequence as base number 1.

Table 4–7 contains the results of digestions of YIP5 DNA using several different combinations of restriction enzymes. The sizes of the DNA fragments produced in the digestions are given in base pairs.

Table 4-7. Fragment Sizes from Digestion of YIP5 DNA

Restriction Enzymes in Digest	Fragment Sizes (Base Pairs)
Eco R1	5541
Hind III	5541
Pst I	2390, 3151
Eco R1 + Hind III	32, 5509
Eco R1 + Pst I	751, 1639, 3151
Eco R1 + Pst I + Hind III	32, 751, 1607, 3151

From the digests using one restriction enzyme, we are able to determine the number of restriction sites for each enzyme. The Eco R1 and Hind III enzymes have just one recognition site, but Pst I has two sites, as shown by the two fragments produced by digestion with Pst I alone.

The positions of the Hind III and Pst I sites relative to the Eco R1 site are determined from digests with different combinations of enzymes. The Hind III site is 32 bp away from the Eco R1 site, but in which direction around the circle is only determined by using another enzyme in the combination digest, such as Pst I. Pst I has two restriction sites and the Eco R1 site is in the 2390-bp fragment. The Hind III site is in the 1639-bp fragment produced by cleavage with both Eco R1 and Pst I. Ordering the fragments in a circle, we obtain the diagram of YIP5 DNA shown in Fig. 4–11.

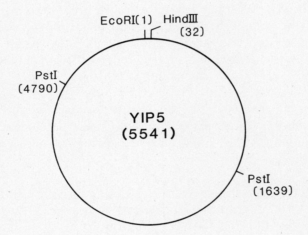

Figure 4–11.

Problem 4–23

The plasmid pUC19, a double-stranded DNA circle of 2686 bp, is used as a cloning vector in *E. coli*. The restriction sites of the enzymes Eco R1, Afl III, and Aha II were mapped from the following data.

Restriction Enzymes in Digests	Fragment Sizes
Eco R1	2686
Afl III	2686
Aha II	304, 382, 2000
Eco R1 + Afl III	806, 1880
Eco R1 + Aha II	69, 235, 382, 2000
Afl III + Aha II	304, 382, 571, 1429
Eco R1 + Afl III + Aha II	69, 235, 382, 571, 1429

(a) How many recognition sites does each enzyme have in the pUC19 plasmid?

(b) Diagram the pUC19 plasmid from the data given below, setting the Eco R1 site at base 1.

Solution:

(a) Eco R1 and Afl III: each have one site in pUC19 as shown by the single DNA fragment in each digest. Aha II has three sites.

(b) To order the fragments, Eco R1 is arbitrarily chosen as cleaving at base 1. Then the Afl III site is at base 806 or 1880. The 304-bp fragment produced by Aha II digests contains the Eco R1 site, indicating that Aha II sites lie on either side of base 1. The Afl III site is in the 2000-bp fragment found in Aha II digests. Figure 4–12 illustrates the most plausible arrangement of restriction sites in pUC19.

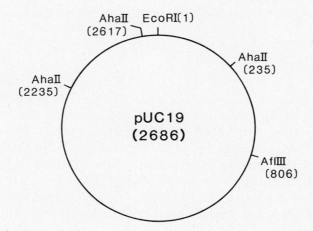

Figure 4–12.

Problem 4–24

Normal human hemoglobin (Hb A) is produced only in red blood cells and is composed of two alpha and two beta subunits. Sickle-cell anemia is an inherited trait in which there are two mutant alleles of the gene coding for the beta subunit and no normal beta subunits are made in the red blood cell, only Hb S, the sickle-cell hemoglobin. The N-terminal sequence of the normal beta subunit was determined to be

```
         1  2  3  4  5  6  7  8
Hb A N   val-his-leu-thr-pro-glu-glu-lys-  beta subunit
```

In tryptic digests of purified Hb S beta subunit, only one peptide fragment was different. This fragment turned out to be the N terminal of the beta subunit, which differed from the normal by only one residue:

```
         1  2  3  4  5  6  7  8
Hb S N   val-his-leu-thr-pro-val-glu-lys-  beta subunit
```

What single base change in the antisense strand of the DNA coding for the beta subunit would cause the sickle-cell hemoglobin, Hb S?

Solution:

The codon dictionary (Fig. 4–10) gives the three-base code in the mRNA for each amino acid.

Codons for val: GUA, GUG, GUC, and GUU

Codons for glu: GAA and GAG

Therefore, the most likely single base change in the DNA would be from A in the antisense strand of the normal gene to T in the mutant allele for sickle-cell beta subunit.

Problem 4–25

The accurate detection of sickle-cell anemia in human fetuses has been extremely difficult until restriction enzyme digests of DNA from fetal cells were shown to be reliable in determining the presence of the Hb S allele. The restriction enzyme, Mst II, recognizes the DNA base sequence CCT/GAG, cleaving the DNA between the T and the G. There is a CCTGAG site in the DNA of the normal beta-globin allele at the fifth and sixth codons. In the sickle-cell allele, the A-to-T substitution eliminates the Mst II site.

In Mst II restriction digests of the DNA from normal individuals, a fragment of 1.1 kb is known to carry the beta-globin gene. However, in the Mst II restriction digests of DNA from persons with sickle-cell anemia, the 1.1-kb fragment is missing. Instead only a 1.3-kb fragment is found which contains the beta globin for Hb S.

Millions of people are carriers of the sickle-cell gene; that is, they are heterozygous for the mutant allele and under normal circumstances do not suffer from the disease. From the diagrams of Mst II digests in Fig. 4–13, identify the DNA from a normal person, a sickle-cell carrier, and a person homozygous for sickle-cell anemia.

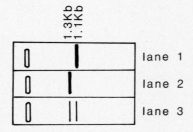

Figure 4–13. Mst II digests of DNA from people with Hb A, Hb S, and Hb A/Hb S genotypes.

Solution:

In Fig. 4–13, lane 1 shows the 1.1-kb fragment found in digests of normal Hb A DNA. Lane 2 shows the 1.3-kb fragment characteristic of Hb S DNA. Lane 3 shows the results from an Mst II digest of DNA from a carrier of sickle-cell anemia (Hb A/Hb S) showing both sizes of fragments, 1.1 kb and 1.3 kb.

DNA Sequencing

There are two methods that have been used to determine the sequence of bases in fragments of DNA, the Maxam and Gilbert method, and the dideoxy or chain-termination method developed by Sanger and his colleagues.

Maxam and Gilbert Method. Maxam and Gilbert developed a technique for determining the base sequence in single-strand fragments of DNA by selectively destroying the 3′-5′ phosphoester backbone between specific nucleotides and then determining the size of the remaining fragments.

Long strands of double-stranded DNA are first cleaved with restriction enzymes to give smaller dsDNA fragments which are heated to produce single strands of DNA(ssDNA). The ssDNA fragments are treated with a phosphatase enzyme to remove the 5′-P and then with nucleotide kinase, which adds radioactively labeled PO_4 to the 5′ end.

Aliquots of 5′-P-labeled ssDNA fragments are subjected to chemical treatments which selectively alter specific bases. Subsequent treatment with piperadine breaks the 3′,5′-phosphoester bonds. These chemical digestions are described and the results shown in Table 4–8.

Note: Each treatment is controlled so that only about 30% of the possible bonds are cleaved.

Table 4–8. Selective Chemical Digestions of ssDNA (5′-p-*AATTCG-3′)[a]

Nucleotide Cleavages	Reagents Used	Fragments Produced
Pyrimidines		
Thymine and cytosine	Hydrazine	*AA; *AAT; *AATT
Cytosine	Hydrazine + 5 *M* NaCl	*AATT
Purines		
Guanine and adenine	Piperidine and formate, pH 2.0	*A; *AATTC
Guanine	DMS (dimethyl SO_4), pH 3.0; piperidine	*AATTC

[a]Only the 5′ radioactively labeled fragments may be detected.

All the fragments produced in the chemical digests are separated on agarose gels by electrophoresis. The gels are pressed against X-ray film so that the DNA fragments containing radioactively labeled fragments may be detected after the film is developed. A diagram of the film is shown in Fig. 4–14.

Problem 4-26

Figure 4–14 shows the results of sequencing a portion of a promoter region found in bacterial DNA. From this diagram, determine the base sequence in this DNA fragment.

Solution:

Reading from the bottom, base 1 (5′ end) is T. The sequence is

5′-TCGTCAAGAATTAATC

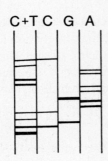

Figure 4–14.

The Dideoxy Procedure of Sanger et al. The dideoxy procedure developed by Sanger and his colleagues has become more popular, partly because the hazardous chemicals used in the Maxam and Gilbert method are not required. In this procedure the DNA strand to be sequenced is used as a template for the synthesis of the complementary strand in the presence of:

1.   Deoxynucleotide triphosphate substrates, dGTP, dCTP, and dTTP

2.   The dATP substrate radioactively labeled with 32P

3.   The Klenow fragment, a portion of the DNA polymerase I enzyme which catalyzes the synthesis of DNA

4.   A dideoxynucleoside triphosphate which terminates the growing DNA chain when incorporated into it (e.g., ddGTP would stop the synthesis at each G added to the chain)

The partially synthesized DNA fragments are separated by electrophoresis. The fragments are located by autoradiography since one of the nucleotide substrates is radioactively labeled. The base sequence is determined by the size of the fragments with the smallest fragments at the bottom of the gel corresponding to the 5′ end of the growing chain.

In the example below, a single strand of DNA is sequenced using the dideoxy method. The base sequence in the strand is 5′-TAGCTAATGC-3′. Since the complementary strand is always antiparallel and is always synthesized starting at the 5′ end, the nucleotide that initiates the synthesis will be dGTP, followed by dCTP and then dATP, which carries the radioactive label.

        5′ TAGCTAATGC 3′

        3′   CG 5′

            dATP*

If dideoxy TTP (ddTTP) is the chain-terminating nucleotide in this experiment, the synthesis will end after the addition of ddTTP.

        5′ TAGCTAATGC 3′

        3′   TA*CG 5′

The products of the DNA synthesis are separated by electrophoresis on agarose gels.

In the next experiment, dideoxy GTP may be used and the synthesis of the complementary strand would be terminated after the addition of each ddGTP, resulting in a one nucleotide fragment with no radioactive label. This fragment would not be detected because it is not radioactive, having no chance to add dATP* to the growing chain.

Problem 4–27

The DNA fragment sequenced in Fig. 4–15 was obtained from restriction digests of DNA purified from rice chloroplasts by the dideoxy method. From the data given, deduce the the sequence of bases in this fragment of DNA.

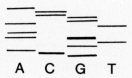

Figure 4–15. The letter below each lane of the gel indicates the dideoxy nucleotide that terminated the growing chain.

Solution:

The sequence is

5'-GCAGTGAATAGGCCA-3'

Problem 4-28

In the chain termination method of DNA sequencing, ddGTP is added to stop the growing DNA chain after G has been added. Why is dGTP, the normal nucleotide, still added to the synthesis reaction?

Solution:

The normal dGTP is added together with the ddGTP so that only about 10% of the chains will be terminated at the first G residue added. Most growing chains will not be terminated until G is added farther along the DNA sequence. This allows the sequence determination of DNA fragments up to 70 to 100 nucleotides in length.

Exercises

The answers are provided at the end of the book.

4-1. The enzyme histidine decarboxylase catalyzes the removal of the carboxyl group from histidine, producing histamine and CO_2. A suspension of bacterial cells was sonicated and centrifuged. Ammonium sulfate crystals were added to the supernatant to give 50% saturation. The protein precipitate was discarded and AS was added to the supernatant to give 70% AS saturation. The precipitate that formed was dissolved in buffer and heated to 70°C for 2 min. The denatured protein precipitate was removed by centrifugation. An aliquot (1.0 mL) of the supernatant was added to the Sephadex G-200 column in Fig. 4-4 and the enzyme's activity eluted at the end of the void volume.

(a) Summarize the results of the purification scheme in a flowchart.

(b) Calculate the specific activity and fold purification for each step in Table 4-9.

(c) What is the molecular weight of the native histidine decarboxylase?

(d) Pure samples of this enzyme were placed on SDS 11% gels (Fig. 4-6b) and two bands were seen. One band has $R_f = 0.45$ and the other band was very close to the dye front, R_f approximately 0.9. What can you determine about the subunit structure of this enzyme from these data?

Table 4-9. Purification of Histidine Decarboxylase

Step	Total Volume (mL)	Enzyme Activity in 0.1 mL (μmol/min)	Total Protein (mg)	Specific Activity	Fold Purification
Cell extract	178	1.9	1489		
AS ppt. in buffer	25	3.2	203		
After heat treatment	20	3.8	37		
Fractions 40–43 from G-200 column	8	8.2	5.4		

4-2. The enzyme phosphoglucoisomerase (PGI) catalyzes the conversion of glucose-6-PO_4 to fructose-6-PO_4 in the glycolysis pathway. Purified PGI was found to elute from the Sephadex G-200 column (Fig. 4-3) with a peak at $V_e = 49$ mL. Only one band was found after PGI was run on SDS-PAGE (11%) and stained with Coomassie Blue. This band had an R_f value of 0.50. What can you conclude from these data about the structure of PGI?

4-3.   A small peptide (angiotensin I) that appears to influence blood pressure has been isolated and purified from human blood. After acid hydrolysis, the amino acid composition was found to be

1 arg: 1 asp: 2 his: 1 ile: 1 leu: 1 phe: 1 pro: 1 tyr: 1 val

The dansyl chloride derivative was found to be asp. Carboxypeptidase A cleaved leu from the peptide. After digestion of the peptide by trypsin, one large peptide fragment and a dipeptide were isolated. Using the Edman degradation method, the dipeptide was shown to be asp-arg.

Chymotrypsin digestion yielded three peptide fragments, free tyr and ile. Two of these peptide fragments were sequenced by the Edman procedure and had the following sequences:

Chymotrypsin digest fragments:

his-leu   asp-arg-val

Thermolysin digested angiotensin I to produce free leu, val, and tyr; a dipeptide; and the peptide fragments phe-his and ile-his-pro. From the data above, determine the amino acid sequence of this peptide.

4-4.   The plasmid pBR322 is used routinely as a cloning vehicle. It is double-stranded DNA of 4400 base pairs (bp) and exists in _E. coli_ as a circle. Genes for resistance to the antibiotics tetracycline and ampicillin are contained in pBR322. Each of the restriction enzymes given below has only one cleavage point in the pBR322 DNA, and each RE, by itself, converts the pBR322 into a linear molecule. When the restriction enzymes are combined, two fragments of the pBR322 DNA are produced and a map of the restriction sites may be developed. From the size of the DNA fragments given for the digests below, construct a restriction map of the pBR322 plasmid.

Restriction Enzymes Used in Digests	Size of pBR322 Fragments (bp)
Eco R1 + Hind III	29, 4334
Eco R1 + Bsm U	1353, 3010
Eco R1 + Pst I	754, 3609
Hind III + Pst I	783, 3580
Eco R1 + Hind III + Pst I	29, 754, 3580
Eco R1 + Bsm I + Hind III + Pst I	29, 1323, 754, 2256

4-5.   Hexokinase catalyzes the formation of glucose-6-PO_4 from glucose and ATP. In experiments to study the structure of purified hexokinase from fungal cell extracts, the following results were obtained:

Three bands of hexokinase *activity* were seen after polyacrylamide gel electrophoresis.

Two bands were seen on SDS-PAGE gels (7%); one band had $R_f = 0.65$ and the other band had $R_f = 0.68$.

Hexokinase activity was found to elute in a broad peak in fractions 47 to 57 mL from the Sephadex G-200 column (Fig. 4-3).

What can you conclude about the structure of this fungal hexokinase from these data?

4-6.   A peptide purified from bacterial culture media was studied and gave the following data:

After acid hydrolysis, the amino acid composition was found to be

2 ile: 1 asn: 1 gln: 2 leu: 2 pro: 1 tyr: 2 val

No dansyl chloride derivative was obtained.

Carboxypeptidases failed to cleave any amino acid from the intact protein.

No titratable groups were observed in titrations of the intact peptide.

What can you conclude about the structure of this peptide from this data?

Chapter 5
BIOENERGETICS

All cells require energy to remain viable. Energy may be found in our environment in many forms (e.g., heat, pressure, light, electricity, sound, chemical fuels, and radioactivity). Animals can use only the chemical (carbon) fuels in the nutrients they ingest. These chemicals (sugars, fats, and proteins) contain potential energy which is released during the controlled degradation of nutrients in cells. The energy released is harnassed by the cell and stored in the high-energy bonds in the nucleoside triphosphate, ATP. Solar energy in the form of light is used by plants with chlorophyll pigments to form energy-rich compounds.

The energy derived from our environment is used by cells to maintain chemical gradients across cellular membranes, produce electrical impulses (transmission of nervous impulses), achieve mechanical work (muscle contraction) and synthesize molecules such as proteins, complex lipids, polysaccharides, and nucleic acids.

In studying the energy relationships in biological systems, the concepts and quantitative methods of thermodynamics are used. These concepts and equations were developed by studying individual chemical reactions *outside the cell, at equilibrium, under standard conditions where the temperature is constant (25°C) and the concentrations of reactants are 1.0 M.* It is important to understand that in cells, an individual chemical reaction is part of a sequence of enzyme-catalyzed reactions and is never isolated. Also, chemical reactions are rarely at equilibrium in the cell and *standard conditions are never found in biological systems.* Nevertheless, the models and concepts that have been developed are valuable, and the quantitative methods allow us to approximate the energy relationships in cells.

ΔG_f° Is the Standard Free Energy of Formation

The ΔG_f° for a given compound is that amount of energy which is required to form 1.0 mol of that compound at 25°C from its elements; for example, the ΔG_f° for glucose = −219 kcal/mol.

The ΔG_f° is different for each different molecule (Table 5–1). If a compound is changed, structurally, during a chemical reaction, the resulting product will have a different ΔG_f°. The difference between the ΔG_f°'s of the product and the reactant is the standard free energy change of the reaction, ΔG°.

Table 5-1. ΔG_f° Values of Some Biochemical Compounds.

Compound	G_f° (kcal/mol)
L-Alanine	−88.4
Glucose	−219
Citrate	−279.24
Isocitrate	−278.47
Water	−56.7
Fumarate	−142.8
Malate	−200.18
Glycine	−88.2
Alanylglycine	−117.1

ΔG, Free Energy Change

Free energy (ΔG) is the potential energy that may be released from a molecule to do useful work. Useful work in the cell usually involves forming or breaking chemical bonds, or maintaining concentration gradients across membranes. For example, the free energy released by electrons moving from one cytochrome to another is harnessed and used to maintain a proton gradient which makes it possible for ATP to be formed. In most biochemical reactions that occur in cells, there is a decrease in free energy ($-\Delta G$) from reactants to products.

ΔG° Is the Change in Standard Free Energy of a Chemical Reaction

ΔG° is the *standard free energy change* and equals the *free energy (ΔG) released in a reaction under standard conditions.* Although standard conditions never exist in cells and reactions do not come to equilibrium in cells, the ΔG° is a useful term because it allows us to compare the energy released in different reactions under identical conditions.

The term $\Delta G^{\circ\prime}$ is the standard free energy change under standard conditions at pH 7.0. $\Delta G^{\circ\prime}$ may vary considerably from ΔG°.

The ΔG° is also a function of the equilibrium constant, K_{eq} for the reaction. The ΔG° for a given reaction may be determined by the following expressions:

(1) $\Delta G^{\circ} = \Delta G_f^{\circ}$ of products $- \Delta G_f^{\circ}$ of reactants

(2) $\Delta G^{\circ} = -1.363 \log K_{eq}$

If the ΔG° value is negative, the reaction is exergonic and may proceed under favorable conditions. If the ΔG° is positive, the reaction is endergonic and will *not* proceed under standard conditions but may proceed under certain cellular conditions.

Problem 5-1

Calculate the ΔG° for each of the following reactions from the ΔG_f° values in Table 5-1 Assume that the product is the compound on the right in each reaction.

(a) Citrate $\rightleftharpoons$ isocitrate

(b) Alanylglycine $\xrightarrow[H_2O]{}$ L-ala + gly

Solution:

(a) $\Delta G^\circ = (-278.47) - (-279.24)$

$\qquad = -278.47 + 279.24$

$\qquad = +0.77 \text{ kcal/mol}$

(b) $\Delta G^\circ = (-88.2 + -88.4) - (-117.24 + -56.7)$

$\qquad = -176.6 - (-173.9)$

$\qquad = -176.6 + 173.9$

$\qquad = -2.7 \text{ kcal/mol}$

Problem 5-2

Determine the ΔG° for the following reactions from their K_{eq} values.

(a) Glucose-6-P $\rightleftharpoons$ fructose-6-P; $K_{eq} = 0.509$

(b) ATP $\longrightarrow$ ADP + HPO_4^{2-} ; $K_{eq} = 9.2 \times 10^5$

Solution:

(a) Since $\Delta G^\circ = -1.363 \log K_{eq}$,

$\qquad \Delta G^\circ = -1.363 \log (0.509) = -1.363 \, (-0.293)$

$\qquad\quad = +0.399 \text{ kcal/mol}$

(b) $\Delta G^\circ = -1.363 \log (9.2 \times 10^5) = -1.363 \, (5.96)$

$\qquad\quad = -8.13 \text{ kcal/mol}$

Relating ΔG° to ΔG

The change in free energy (ΔG) in a biochemical reaction is directly related to (1) the ΔG° value, (standard free energy change) and (2) the concentrations of the product(s) and reactant(s).

In the reversible reaction

$$A + B \rightleftharpoons C + D \tag{1}$$

A and B may be the reactants or the products, depending on the *direction* in which the reaction goes.

Note: In the cell, reactions always occur in the direction giving *negative* ΔG values.

To determine the ΔG value for a reaction, the following equation is used:

$$\Delta G = \Delta G^\circ + RT \ln K_{eq} \tag{2}$$

where R is the gas constant, 1.98 cal/mol or 0.00198 kcal/mol; T the absolute temperature, 298 K (25°C); ln the natural log, base e; and K_{eq} the equilibrium constant. In the reaction $A + B \rightleftharpoons C + D$,

$$K_{eq} = \frac{[C][D]}{[A][B]}$$

Equation (2) may be rewritten

$$\Delta G = \Delta G^\circ + 0.592 \ln \frac{[C][D]}{[A][B]} \tag{3}$$

when $R = 0.00198$ kcal/mol.

We may convert ln to log, base 10, by multiplying the expression by 2.303 (see Chapter 1). Rewriting equation (3) using $\log_{10}$, we have

$$\Delta G = \Delta G^\circ + 1.363 \log_{10} \frac{[C][D]}{[A][B]} \tag{4}$$

Equation (4) is most useful in determining the *direction* of a reaction and the amount of energy released if the reaction proceeds. It tells us nothing about the *rate* at which the reaction occurs. Without an enzyme to catalyze it, virtually no reaction will proceed in the cell, no matter how favorable the ΔG.

Note: The *sign* of ΔG indicates the *direction* of the reaction. The *rate of the reaction is determined by the amount of enzyme and the kinetic constants, K_m, V_{max}, of the enzyme that catalyzes it.*
         NO ENZYME CAN CHANGE THE DIRECTION OF THE REACTION.

Problem 5-3

(a) From the ΔG_f° values below, calculate ΔG° (standard free energy change) for the reaction

$$\text{glucose-6-PO}_4 \rightleftharpoons \text{glucose-1-PO}_4$$

with glucose-1-PO$_4$ as product.

ΔG_f° for G-6-P $= -421.1$ kcal/mol

ΔG_f° for G-1-P $= -421.0$ kcal/mol

(b) What is the ΔG° for the reverse reaction?

(c) If [glucose-1-PO$_4$] = [glucose-6-PO$_4$], what will the free energy change be?

(d) What can you conclude about the *rate* of the reaction from parts (a), (b), and (c)?

Solution:

(a) ΔG° for the reaction with G-1-P as product is

$$\Delta G^{\circ} = (-421.0) - (-421.1)$$

$$= +0.1 \text{ kcal/mol}$$

(b) ΔG° for the reverse reaction is -0.1 kcal/mol.

(c) If [P] = [R], then $\Delta G = \Delta G^{\circ}$, since [P]/[R] = 1.0 and the log of 1 = 0. The term 1.363 log [P]/[R] then becomes zero.

(d) You can conclude *nothing* about the *rate* of the reaction from these data, since the rate of the reaction depends on the enzyme that catalyzes it.

Problem 5–4

In the tricarboxylic acid (TCA) cycle, the enzyme aconitase catalyzes the removal of a H$_2$O molecule from citrate to give the product *cis*-aconitate. This reaction is reversible in the cell. What concentration of citrate is required if the reaction is to proceed toward *cis*-aconitate as product? Assume that [*cis*-aconitate] = 10^{-4} M.

Solution:

$$\Delta G^{\circ} = \Delta G_f^{\circ} \text{ of products} - \Delta G_f^{\circ} \text{ of reactants}$$

$$\Delta G^{\circ} = [(-220.5) + (-56.7)] - (-279.24)$$

$$= -220 - 56.7 + 279.24$$

$$= +2.0 \text{ kcal/mol}$$

Thus

$$\Delta G = +2.0 + 1.363 \log_{10} \frac{[\textit{cis}\text{-aconitate}][H_2O]}{[\text{citrate}]}$$

Note: The standard free energy of formation for H_2O is used to calculate $\Delta G°$ for the reaction. The cell is an aqueous medium, but the [free H_2O] is not known. For the purposes of problem solving, we assume that $[H_2O] = 1\ M$. Log[free H_2O] is log 1, which is zero. Thus the equation above simplifies to

$$\Delta G = +2.0 + 1.363 \log_{10} \frac{[10^{-4}]}{[\text{citrate}]}$$

Let $[\text{citrate}] = 10^{-x}\ M$ since, in the cell, the concentrations of metabolites such as citrate are always less than $1\ M$. Then

$$\Delta G + 2.0 + 1.363 \log_{10} \frac{10^{-4}\ M}{10^{-x}\ M} \qquad \text{(see Chapter 1, Problem 1–20(c))}$$

Simplifying gives

$$\Delta G = +2.0 + 1.363(-4 + x)$$

$$= +2.0 - 5.45 + 1.363x$$

For the reaction to proceed toward *cis*-aconitate, the ΔG must be less than 0. If we set $\Delta G = 0$ (equilibrium conditions), we can solve for x and determine the [citrate] at equilibrium. *If [citrate] increases above this equilibrium value, the reaction will proceed toward cis-aconitate.* If [citrate] decreases to below the equilibrium value, the direction of the reaction will be toward citrate as product.

$$0 = +2.0 - 5.45 + 1.363x$$

$$1.363x = 3.45$$

$$x = 2.53$$

Then the [citrate] at equilibrium is $10^{-2.53}\ M$. Taking the antilog of $10^{-2.53}$, we obtain 0.00295. Therefore, the [citrate] must exceed $0.00295\ M$ or 2.95 mM for this reaction to proceed toward *cis*-aconitate as the product.

Problem 5-5

Aldolase is the enzyme in glycolysis which cleaves fructose-1,6-bisPO$_4$ to yield two three-carbon PO$_4$ molecules, 3-phosphoglyceraldehyde (3-PG) and dihydroxyacetone phosphase (DHAP).

What is the lowest concentration of F-1,6-bisPO$_4$ which will allow this reaction to proceed toward 3-PG and DHAP as products, given that

$\Delta G^{\circ\prime}$ for the cleavage reaction = +5.73 kcal/mol

$[\text{3-PG}] = 10^{-5.2}\ M$

$[\text{DHAP}] = 10^{-3.8}\ M$

$$\text{F-1,6-bisP} \rightleftharpoons \text{DHAP} + \text{3-PG}$$

Solution:

Let [F-1,6-bisP] be $10^{-x}\ M$. Substituting in equation (4) gives us

$$\Delta G = \Delta G^{\circ\prime} + 1.363 \log \frac{[\text{3-PG}][\text{DHAP}]}{[\text{F-1,6-bisPO}_4]}$$

$$= +5.73 + 1.363 \log \frac{10^{-5.2} \times 10^{-3.8}}{10^{-x}}$$

$$= +5.73 + 1.363(-5.2 - 3.8 + x) = +5.73 + 1.363(-9.0 + x)$$

If we assume equilibrium conditions, we may set $\Delta G = 0$. Then

$$-5.73 = +1.363(-9.0 + x)$$

$$-5.73 = -12.3 + 1.363x$$

$$1.363x = 6.57$$

$$x = 4.82$$

At equilibrium, [F-1,6-bisPO$_4$] = $10^{-4.82}\ M$ or $1.51 \times 10^{-5}\ M$. In order for the reaction to proceed toward 3-PG and DHAP as products, [F-1,6-bisP] must be greater than $1.51 \times 10^{-5}\ M$.

Problem 5-6

The enzyme phosphoglucose isomerase catalyzes the formation of glucose-1-PO$_4$ from glucose-6-PO$_4$. Find the [G-1-P]/[G-6-P] ratio that would allow the phosphoglucose isomerase reaction to go toward glucose-1-PO$_4$ as the product, assuming that the $\Delta G^{\circ\prime} = +0.1$ kcal/mol for glucose-1-P as product, and [G-1-P] = 0.4 mM.

Solution:

Let the ratio of [G-1-P]/[G-6-P] = x.

$$\Delta G = \Delta G^{\circ\prime} + 1.363 \log (x)$$
$$= +0.1 + 1.363 \log (x)$$

At equilibrium, $\Delta G = 0$. Therefore, at equilibrium,

$$0 = +0.1 + 1.363 \log (x)$$

$$\frac{-0.1}{1.363} = \log (x)$$

$$-0.073 \doteq \log (x)$$

$$\text{antilog} (-0.073) = 0.84 = x$$

Then, at equilibrium, the ratio [G-1-P]/[G-6-P] equals 0.84. For the reaction to proceed toward G-1-P, the ratio must be less than 0.84.

Problem 5-7

If [G-1-P] were 0.4 mM, what [G-6-P] would be required for the reaction to go toward G-1-P as product in Problem 5-6.

Solution:

If the ratio must be less than 0.84, then substituting [G-1-P] = 0.4 mM, we have

$$\frac{0.4}{[\text{G-6-P}]} = 0.84 \qquad [\text{G-6-P}] = \frac{0.4}{0.84} = 0.476$$

Then the concentration of G-6-P must exceed 0.476 mM.

Direction of Reactions in Cells

Reactions occur only in the direction of $-\Delta G$ values in cells. To determine ΔG for the reverse reaction, we use equation (4) with the opposite sign for ΔG° and the concentrations of former products as the reactants. Some reactions are easily reversed in the cell simply by decreasing the [P] and increasing the [R] slightly. These easily reversed reactions generally have very small ΔG° values (less than 4 kcal/mol), either + or − in sign.

When a reaction is reversed, the ΔG° value remains the same but the sign is changed. In a reaction with a large $-\Delta G^{\circ}$, the reverse reaction would have a large $+\Delta G^{\circ}$ value. Since $\Delta G = \Delta G^{\circ} + 1.363 \log [\text{P}]/[\text{R}]$, then if the [P] was very close to zero and the [R] was quite large, the term [P]/[R] would be a very small fraction and the log of the term would be a very large negative value. This would overcome many large $+\Delta G°$ values and the reaction could reverse. However, in the cell it may be impossible to achieve the extremely large changes in concentrations of the reactants and products required to reverse the reaction.

In the cell, reactions are *not* isolated as they may be in a test tube and the concentration of reactants and products are controlled by other reactions in other metabolic pathways. Therefore, *many reactions that are reversible in vitro are not reversible in the cell.*

Problem 5-8

What is the $\Delta G°'$ for the reactions below proceeding in the *reverse* direction?

$\Delta G°'$ in Direction Shown (kcal/mol)	Reaction
−3.4	F-6-P + ATP → F-1,6-bisP + ADP
+6.0	Lactate + NAD^+ → pyruvate + NAD·H
+5.73	F-1,6-bisP → 3PG + DHAP
−0.88	Fumarate + H_2O → malate
0.0	Succinate + FAD → fumarate + FAD·H_2

Solution:

$\Delta G°'$ (kcal/mol)	Reaction
+3.4	F-1,6-bis + ADP → F-6-P + ATP
−6.0	Pyruvate + NAD·H → lactate + NAD^+
−5.73	3PG + DHAP → F-1,6-bisP
+0.88	Malate → fumarate + H_2O
0.0	Fumarate + FAD·H_2 → succinate + FAD

Problem 5-9

In what direction will the following reaction proceed under the conditions given below?

oxaloacetate + NAD·H $\rightleftharpoons$ malate + NAD^+
    (OAA)

$\Delta G°'$ = −6.7 kcal/mol for malate and NAD^+ as products

[OAA] = 0.1 μM

[malate] = $2 \times 10^{-5} M$

$$[NAD^+] \ = 0.3 \text{ m}M$$

$$[NAD \cdot H] \ = 0.3 \text{ m}M$$

Solution:

Determine the ΔG in the direction for malate and NAD^+ as products. Substituting in the equation $\Delta G = \Delta G^{\circ\prime} + 1.363 \log [P]/[R]$ gives us

$$\Delta G = -6.7 + 1.363 \log \frac{10^{-4.7} \times 10^{-3.52} \ M}{10^{-7} \times 10^{-3.52} \ M}$$

$$= -6.7 + 1.363 \ (-8.22 + 10.52)$$

$$= -6.7 + 3.13$$

$$= -3.6 \text{ kcal/mol}$$

Since the ΔG is negative, the reaction would proceed toward malate and NAD^+ as products under these conditions.

Problem 5-10

What concentration of oxaloacetate (OAA) would be required to cause the reaction above to proceed toward OAA as product if the [malate] increased to $8 \times 10^{-4} \ M$?

Solution:

Let the [OAA] be $10^{-x} \ M$ and set $\Delta G = 0$ (equilibrium conditions). Then solving for x should allow us to determine the [OAA] required for equilibrium conditions. Since the reverse reaction is being considered, we change the sign of $\Delta G^{\circ\prime}$ above and now $\Delta G^{\circ} = +6.7$ kcal/mol. Substituting in the equation $\Delta G = \Delta G^{\circ\prime} + 1.363 \log ([P]/[R])$ yields

$$0 \ = +6.7 + 1.363 \log \frac{[OAA][NAD \cdot H]}{[malate][NAD^+]}$$

$$= +6.7 + 1.363 \log \frac{10^{-x} \ M \times 10^{-3.52} \ M}{10^{-3.1} \ M \times 10^{-3.52} \ M}$$

$$= 6.7 + 1.363 \ (-x - 3.52 + 6.62)$$

$$= 6.7 - 1.363x + 4.22$$

$$x \ = \frac{10.92}{1.363}$$

$$= 8.01$$

Then [OAA] must be decreased to less than $10^{-8.0} \ M$ for this reaction to proceed toward OAA and $NAD \cdot H$ as products under these conditions.

Problem 5-11

A key regulatory enzyme in the glycolysis pathway is phosphofructokinase (PFK). The reaction catalyzed by this enzyme is generally considered to be irreversible in cells. The $\Delta G^{\circ\prime}$ is -3.4 kcal/mol with ADP and F-1,6-bisP as products. The approximate range of concentrations for each product and reactant in most cells is shown below. Given these conditions, show why the PFK reaction is *not* reversible in cells.

$$\text{F-6-P} \quad + \quad \text{ATP} \longrightarrow \text{F-1,6-bisP} \quad + \quad \text{ADP}$$
$$(0.02-0.06 \text{ m}M) \quad (1-2 \text{ m}M) \qquad (0.01-0.03 \text{ m}M) \quad (0.5-1.0 \text{ m}M)$$

Solution:

Calculate the ΔG for F-1,6-bisP formation under conditions *most* favorable for *reversal* of the PFK reaction, that is, highest [products] and lowest [reactants]. If the ΔG is still negative under these conditions, the probability of reversing it is very low. If the ΔG is positive, the reaction will go in the opposite direction under these conditions.

$$\Delta G = -3.4 + 1.363 \log \frac{[\text{F-1,6-bisP}][\text{ADP}]}{[\text{F-6-P}][\text{ATP}]}$$

Since the conditions most favorable for reversing the reaction are lowest [F-6-P] and [ATP] and highest [F-1,6-bisP] and [ADP], set [ATP] = 1 mM, [F-6-P] = 0.02 mM, [ADP] = 1.0 mM, and [F-1,6-bisP] = 0.03 mM.

$$\Delta G = -3.4 + 1.363 \log \frac{10^{-4.5} \times 10^{-3}}{10^{-4.7} \times 10^{-3}}$$

$$= -3.4 + 1.363 \, [-7.5 - (-7.7)]$$

$$= -3.4 + 1.363 \, (+0.2)$$

$$= -3.13 \text{ kcal/mol}$$

Thus ΔG is negative, indicating that the reaction will still proceed toward F-1,6-bisP and ATP as products under cellular conditions most favorable for reversal. Therefore, *in the cell,* the PFK reaction is irreversible.

Problem 5-12

Although the reaction catalyzed by PFK is apparently irreversible in cells, it is reversible *in a test tube.* Which of the following changes from the conditions in the problem above would allow PFK to catalyze the formation of F-6-P and ATP from the reactants F-1,6-bisP and ADP *in vitro.* (Assume that $\Delta G^{\circ\prime} = +3.4$ kcal/mol for F-6-P + ATP formation.)

(a) Raising the [F-1,6-bisP] to 4.0 mM and [ADP] to 2.1 mM

(b) Decreasing the [F-1,6-bisP] to 1 μM and [ADP] to 2.0 μM

(c) Raising the [F-6-P] to 5.0 mM and [ATP] to 5.0 mM

(d) Decreasing the [F-1,6-bisP] to 1 μM and [ADP] to 2 μM

Solution:

Using the equation

$$\Delta G = \Delta G^{\circ} + 1.363 \log \frac{[\text{F-6-P}]\,[\text{ATP}]}{[\text{F-1,6-bisP}]\,[\text{ADP}]}$$

we may determine the sign of ΔG for each of the four conditions. The $\Delta G^{\circ\prime}$ is +3.4 kcal/mol in the direction with ATP as product.

(a) $\Delta G = \Delta G^{\circ} + 1.363 \log \dfrac{10^{-4.7} \times 10^{-3}}{(4.0 \text{ m}M)(2.1 \text{ m}M)}$

$\qquad = +3.4 + 1.363 \, [-4.7 + (-3) - (-2.39 + (2.67)]$

$\qquad = +3.4 + 1.363 \, (-7.7 + 5.06)$

$\qquad = +3.4 - 3.59$

$\qquad = -0.19 \text{ kcal/mol}$

ΔG is negative and the reaction will proceed toward F-6-P under these conditions.

(b) $\Delta G = +3.4 + 1.363 \log \dfrac{[\text{F-6-P}]\,[\text{ATP}]}{10^{-6} \times 10^{-5.69}}$

$\qquad = +3.4 + 1.363 \, [-4.7 - 3 - (-6 + -5.69)]$

$\qquad = +3.4 + 1.363 \, (+3.99)$

$\qquad = +3.4 + 5.4 \text{ kcal/mol}$

Clearly, ΔG will be positive and the reaction will not proceed toward F-6-P under these conditions.

(c) $\Delta G = +3.4 + 1.363 \log \dfrac{10^{-2.3} \times 10^{-2.3}}{10^{-4.5} \times 10^{-3}}$

$\qquad = +3.4 + 1.363 \, (+2.9)$

$\qquad = +3.4 + 3.95$

$\qquad = +7.35 \text{ kcal/mol}$

Since ΔG is +7.35, these conditions do not allow reversal of PFK.

(d) Decreasing the reactants cannot cause the reaction to proceed.

Problem 5–13

Given the following range of concentrations of acetyl CoA, OAA, citrate, and CoA, in mitochondria, is the citrate synthase reaction reversible?

$$\text{acetyl-CoA} + \text{oxaloacetate (OAA)} \longrightarrow \text{citrate} + \text{CoA}$$

(That is, can this reaction go toward acetyl-CoA and OAA as well as toward citrate and CoA?)

The $\Delta G^{\circ\prime}$ for citrate formation is -8.0 kcal/mol. The range for [acetyl-CoA] is 20 to 50 μM; [CoA] varies from 70 to 100 μM; the [OAA] is *estimated* to be between 10^{-6} M and 10^{-5} M and [citrate] between 200 and 300 μM.

Solution:

Calculate the ΔG for formation of citrate under conditions *most favorable for the reverse reaction*, that is, the lowest [reactants] and highest [products]. Then

$$\Delta G = -8.0 + 1.363 \log \frac{[\text{CoA}][\text{citrate}]}{[\text{acetyl-CoA}][\text{OAA}]}$$

$$= -8.0 + 1.363 \log \frac{(300\ \mu M)(100\ \mu M)}{(20\ \mu M)(1\ \mu M)}$$

$$= -8.0 + 1.363 \log \frac{10^{-3.5} \times 10^{-4}}{10^{-4.7} \times 10^{-6}}$$

$$= -8.0 + 1.363 \,[-7.5 - (-10.7)]$$

$$= -8.0 + 4.36$$

$$= -3.65 \text{ kcal/mol}$$

Since ΔG dictates that the reaction will go toward citrate even under most favorable conditions in the cell for reversal, we may conclude that this reaction is irreversible *in the cell.*

Problem 5–14

Many of the enzymes that act to degrade glucose in the glycolysis pathway also catalyze the reverse reactions and allow the formation of glucose (gluconeogenesis) when conditions in the cell change. If the cell has high [ADP] and [1,3-bisPGA], 3-PGA and ATP will be formed (glycolysis). When the [3-PGA] and [ATP] increase and [1,3-bisPGA] and [ADP] decrease, the reaction reverses and 1,3-bisPGA forms in the direction of gluconeogenesis.

$$\Delta G^{\circ\prime} = -4.5 \text{ kcal/mol; glycolysis}$$
$$\text{1,3-bisPGA} + \text{ADP} \underset{\text{gluconeogenesis}}{\overset{}{\rightleftharpoons}} \text{3-PGA} + \text{ATP}$$

In experiments to determine concentrations of compounds in the livers of rats in different nutritional states, the data shown in Table 5–2 were obtained. Given these data, what [1,3-bisphosphoglycerate] is required for:

(a) Glycolysis to proceed in cells from animals starved and refed high carbohydrate?

(b) Gluconeogenesis to proceed in cells from animals starved and refed high carbohydrate?

(c) In which direction would this reaction proceed in the control cells if [1,3-bisPGA] = 0.4 μM?

Table 5–2. Estimated Concentrations of Metabolites (μM) in cells of Rat Liver

Metabolite	Control	Starved	Starved and Refed High Carbohydrate
ATP	1950	1700	2100
ADP	900	1000	600
3-PGA	250	160	200

Solution:

(a) To determine the [1,3-diPGA] required for glycolysis to proceed in starved cells, the ΔG is set at zero and [1,3-diPGA] = 10^{-x} in the following equation.

$$\Delta G = \Delta G^{\circ\prime} + 1.363 \log \frac{[\text{3-PGA}][\text{ATP}]}{[\text{1,3-bisPGA}][\text{ADP}]}$$

$$0 = -4.5 + 1.363 \log \frac{10^{-3.8} \times 10^{-2.78}}{10^{-x} \times 10^{-3}}$$

$$4.5 = 1.363\,(-6.58 + x + 3)$$

$$= -4.87 + 1.363x$$

$$x = 6.87$$

Therefore, [1,3-bisPGA] must exceed $10^{-6.87}\ M$ or 0.13 μM for glycolysis to proceed under these conditions.

(b) To determine the [1,3-bisPGA] required for gluconeogenesis to occur in starved and refed liver cells, the ΔG is set at 0 and the [1,3-bisPGA] = $10^{-x}\ M$. Since this reaction proceeds toward 1,3-bisPGA in gluconeogenesis, the sign of $\Delta G^{\circ\prime}$ is reversed and $\Delta G^{\circ\prime}$ = +4.5 kcal/mol.

$$\Delta G = +4.5 + 1.363 \log \frac{[\text{1,3-bisPGA}][\text{ADP}]}{[\text{3-PGA}][\text{ATP}]}$$

$$0 = +4.5 + 1.363 \log \frac{(10^{-x})(600 \ \mu M)}{(200 \ \mu M)(2100 \ \mu M)}$$

$$-4.5 = 1.363 \log \frac{10^{-x} \times 10^{-3.22}}{10^{-3.69} \times 10^{-2.68}}$$

$$= 1.363 \ (-x - 3.22 + 6.37)$$

$$1.363x = 8.8$$

$$x = 6.4$$

Therefore, [1,3-bisPGA] must be less than $10^{-6.4}$ M or 0.4 μM for gluconeogenesis to proceed under these conditions.

(c) To determine the direction in which this reaction would proceed in the control cells, calculate ΔG using $\Delta G^\circ = -4.5$ kcal/mol; [1,3-bisPGA] = 0.4 μM and the concentrations given in Table 5–2 for 3-PGA, ADP, and ATP.

$$\Delta G = -4.5 + 1.363 \log \frac{[3\text{-PGA}][ATP]}{[1,3\text{-bisPGA}][ADP]}$$

$$= -4.5 + 1.363 \log \frac{10^{-3.64} \times 10^{-2.71}}{10^{-6.4} \times 10^{-3.04}}$$

$$= -4.5 + 1.363 \ (-6.35 + 9.43)$$

$$= -4.5 + 3.09$$

$$= -1.41 \text{ kcal/mol}$$

Then the reaction in control cells would proceed in the direction of $-\Delta G$, that is, toward 3-PGA as product, which is the glycolytic path.

Problem 5-15

In the reaction

$$1,3\text{-bisPGA} + ADP \rightleftharpoons 3\text{-PGA} + ATP$$

which of the following concentrations of 1,3-bisPGA will allow the reaction to go toward 3-PGA as product under the conditions [ADP] = 1 mM, [ATP] = 1.9 mM, and [3PGA] = 0.2 mM? Assume that $\Delta G^{\circ\prime} = -4.50$ kcal/mol in the direction of 3-PGA as product.

(a) 6 μM

(b) 1.4 × 10^{-8} M

(c) 7.2 × 10^{-9} M

(d) 92 nM

Solution:

Substituting in equation $\Delta G = \Delta G° + 1.363 \log [P]/[R]$, let [1,3-bisPGA] be 10^{-x} M. Setting $\Delta G = 0$ at equilibrium conditions, we obtain

$$0 = -4.5 + 1.363 \log \frac{10^{-3.69} \times 10^{-2.72}}{10^{-x} \times 10^{-3.0}}$$

$$= -4.5 + 1.363 (-3.41 + x)$$

$$= -4.5 - 4.65 + 1.363x$$

$$+9.15 = -9.15 + 1.363x$$

$$x = \frac{9.15}{1.363} = 6.71$$

At equilibrium, the [1,3-bisPGA] will be $10^{-6.71}$ M or 0.19 μM. Therefore, the reaction will proceed toward 3-PGA as product when the [1,3-bisPGA] is greater than 0.19 μM and only answer (a) is correct.

Problem 5–16

In the liver of mammals, lactate is converted to pyruvate by the enzyme lactate dehydrogenase (LDH) to begin the pathway of gluconeogenesis. However, the $\Delta G°$ for pyruvate formation is unfavorable; $\Delta G°' = +6.0$ kcal/mol.

$$\underset{(1.0 \, mM)}{\text{lactate}} + \underset{(0.1 \, mM)}{\text{NAD}^+} \overset{\text{LDH}}{\rightleftharpoons} \underset{(0.1 \, mM)}{\text{pyruvate}} + \underset{(0.1 \, \mu M)}{\text{NAD} \cdot \text{H}} + \underset{(0.1 \, \mu M)}{\text{H}^+}$$

(a) Given the concentrations above, in which direction will this reaction proceed?

(b) Which of the following conditions would cause this reaction to go toward pyruvate as product?

   (i) Adding more of the enzyme, LDH
  (ii) Increasing [lactate] to 2.0 mM
 (iii) Decreasing [pyruvate] to 20.0 μM

Solution:

(a) To determine the direction in which this reaction proceeds, the ΔG must be calculated. Assume that the reaction mixture is well buffered so that no change in pH occurs during the reaction.

$$\Delta G = +6.0 + 1.363 \log \frac{10^{-4} \times 10^{-7} \times 10^{-}}{10^{-3} \times 10^{-4}}$$

$$= +0.55 \text{ kcal/mol}$$

Therefore, the reaction will proceed toward lactate as product.

(b) (i) Increasing the amount of enzyme cannot change the direction of a reaction.

(ii) Increasing the [lactate] to 2.0 mM, we have

$$\Delta G = +6.0 + 1.363 \log \frac{10^{-4} \times 10^{-7}}{10^{2.69} \times 10^{-4}}$$

$$= +6.0 + 1.363 (-11 + 6.69)$$

$$= +0.13 \text{ kcal/mol}$$

Since the ΔG is positive after increasing the [lactate] to 2.0 mM, the reaction would still proceed toward lactate.

(iii) Decreasing the [pyruvate] to 20.0 μM gives us

$$\Delta G = +6.0 + 1.363 \log \frac{10^{-4.69} \times 10^{-7}}{10^{-3} \times 10^{-4}}$$

$$= +6.0 + 1.363 (-11.69 + 7)$$

$$= -0.39 \text{ kcal/mol}$$

Under these conditions ([pyruvate] decreased to 20 μM), the ΔG is negative and the reaction will proceed toward pyruvate.

Coupled Reactions: The Cell's Strategy for Accomplishing Energetically Unfavorable Reactions

Most of the knowledge available about the energetics of biochemical reactions was obtained by studying each reaction individually, in vitro, since it is virtually impossible to study individual reactions in the living cell since reactions are never isolated in cells as they are in test tubes. In the cell, each chemical reaction is part of a series of reactions referred to as a metabolic pathway and *the product of one reaction is the reactant in the next reaction.* Some compounds may be the substrate for two or three different enzymes, each in a different pathway, further complicating the study of an individual reaction in the cell.

A reaction that is energetically unfavorable (i.e., strongly endergonic with a large +$\Delta G°$) may proceed in the cell by being coupled to a *strongly exergonic* (large −$\Delta G°$) *reaction,* through an activated intermediate. Coupled reactions are sequential reactions that may occur on the surface of one enzyme or may be catalyzed be separate enzymes acting sequentially.

Determining $\Delta G°$ for a Sequence of Reactions

The $\Delta G°$ values for each reaction in the pathway may be added to give an overall $\Delta G°$ for the sequence of reactions. For example, in the pathway below, reactant A is converted to end product E through a series of enzyme-catalyzed reactions.

$$\begin{array}{ccccccccc}
& \text{enz 1} & & \text{enz 2} & & \text{enz 3} & & \text{enz 4} & \\
A & \longrightarrow & B & \longrightarrow & C & \longrightarrow & D & \longrightarrow & E \text{ (end product)} \\
\Delta G° = -4 & & \Delta G° = +0.1 & & \Delta G° = +1.1 & & \Delta G° = +2.1 &
\end{array}$$

The $\Delta G°$ for the conversion of A to E is the sum of the $\Delta G°$ values. $\Delta G° = (-4 + 3.3) =$ -0.7 kcal/mol for the conversion of A to E. Thus the formation of E from A is energetically favorable, even though three of the four reactions in the sequence have $+\Delta G°$ values.

Problem 5-17

In the TCA cycle, the oxidation of malate to oxaloacetate is catalyzed by malate dehydrogenase and is energetically unfavorable for OAA formation with a $\Delta G°' = +7.1$ kcal/mol. The product, OAA, is bound immediately by citrate synthase, which catalyzes the condensation of OAA and acetyl CoA to form citrate and initiate the TCA cycle. The $\Delta G°'$ for citrate formation from OAA and acetyl CoA is -8.0 kcal/mol. Calculate the $\Delta G°'$ for the conversion of malate to citrate.

Solution:

To determine the $\Delta G°'$ for the formation of citrate from malate the $\Delta G°$ values for the two reactions are added:

$$\Delta G°' = +7.1 + (-8.0) = -0.9 \text{ kcal/mol}$$

Thus the formation of citrate from malate is exothermic.

High-Energy Bonds and Activated Intermediates

Coupling of reactions frequently is achieved by the use of *high-energy bonds* in *activated intermediates*. Covalent bonds generally release less than 5 kcal/mol of useful energy when hydrolyzed, but a few covalent bonds, termed high-energy bonds, release significantly more than 5.0 kcal/mol. Table 5-3 shows the structure and approximate amount of energy released ($\Delta G°$) upon hydrolysis of some common high-energy bonds found in biological compounds.

Table 5-3. High-Energy Bonds in Selected Biological Compounds (Indicated by Arrows)

Bond Name	Structure	Examples	Approximate $\Delta G°$ for Hydrolysis (kcal/mol)
Thioester	$R_1 \!-\! \overset{\overset{\displaystyle O}{\|}}{C} \!-\! S \!-\! R_2$	Acetyl-CoA	-8
Pyrophosphate	$R \!-\! O \!-\! \overset{\overset{\displaystyle O}{\|}}{\underset{\underset{\displaystyle O^-}{\|}}{P}} \!-\! O \!-\! \overset{\overset{\displaystyle O}{\|}}{\underset{\underset{\displaystyle O^-}{\|}}{P}} \!-\! O^-$	ATP	-8

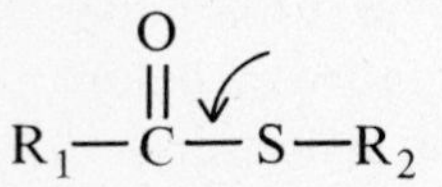

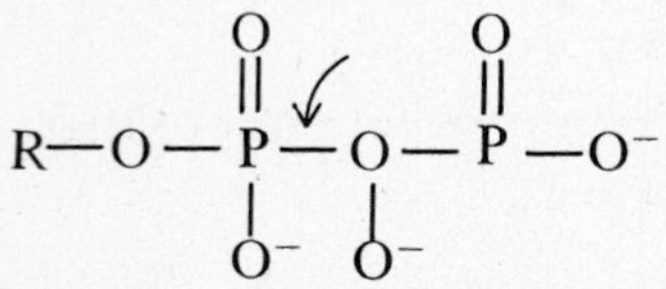

Acyl thiazole	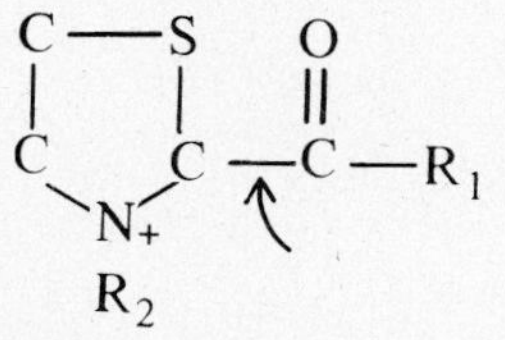	Thiamin pyrophosphate intermediates	-8
Carboxyl PO_4	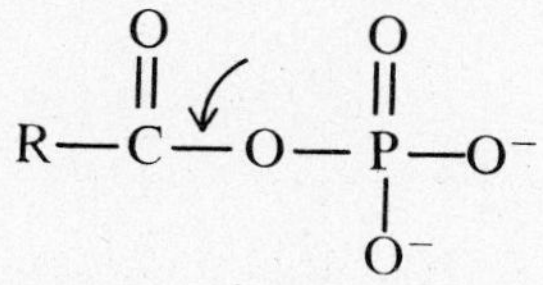	1,3-diphosphoglycerate	-12
Enol PO_4	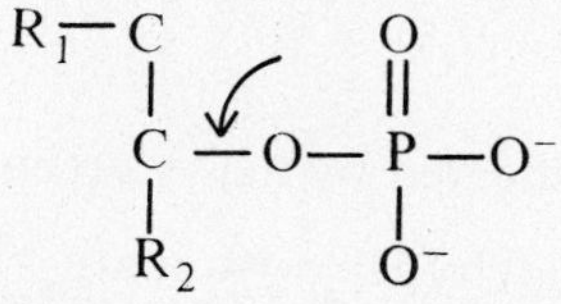	Phosphoenolpyruvate	-15

In many pathways, a strongly exergonic reaction will transfer energy that is stored in a high-energy bond and use that energy to drive an endergonic reaction. For example, the molecule coenzyme A has a free SH group and forms high-energy thioester bonds with many cellular metabolites, such as acetate, succinate, and malonate. Hydrolysis of these bonds releases about 8 kcal/mol ($\Delta G^{\circ\prime}$) and often drives endergonic condensation reactions.

Problem 5–18

In the following reaction sequences, identify the high-energy bonds in the activated intermediates.

(a) pyr $\xrightarrow{}$ thiamin–acetaldehyde $\longrightarrow$ acetaldehyde $\longrightarrow$ lipoate $\longrightarrow$ AcCoa
$\quad\;\; CO_2$

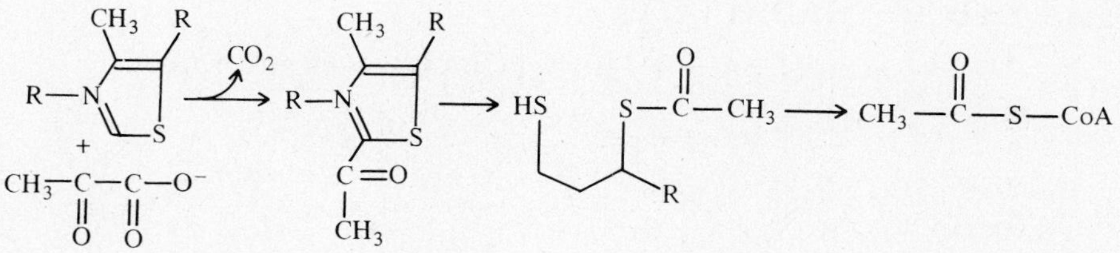

(b)

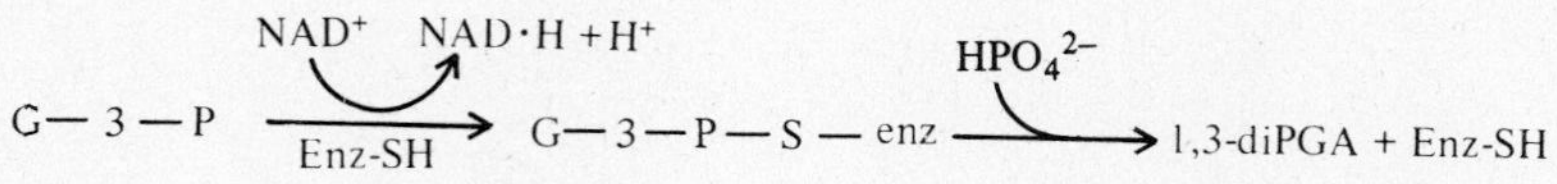

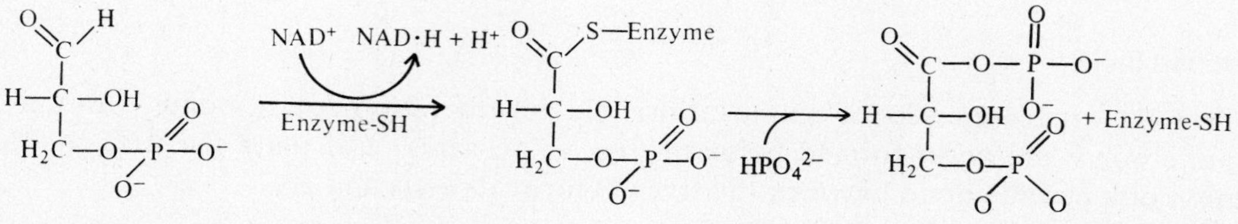

(c) Malate + NAD$^+$ $\longrightarrow$ NAD·H + OAA

acetyl-CoA $\quad$ [citroyl-CoA] $\quad$ citrate + CoA

malate $\qquad$ OAA $\qquad$ citroyl-CoA $\qquad$ citrate

Solution:

The high-energy bonds are shown as ~ and indicated by heavy arrows.

(a) pyr $\xrightarrow{}$ thiamin–acetaldehyde $\longrightarrow$ acetaldehyde $\longrightarrow$ lipoate $\longrightarrow$ AcCoa

(b) $\quad$ G—3—P $\xrightarrow[\text{Enz-SH}]{\text{NAD}^+ \quad \text{NAD·H+H}^+}$ G—3—P—S—enz $\xrightarrow{\text{HPO}_4^{2-}}$ 1,3-diPGA + Enz-SH

(c) Malate + NAD$^+$ $\longrightarrow$ NAD·H + OAA

acetyl-CoA $\longrightarrow$ [citroyl-CoA] $\longrightarrow$ citrate + CoA

malate $\qquad$ OAA $\qquad$ citroyl–CoA $\qquad$ citrate

Problem 5-19

Proteins are composed of amino acids held together by peptide bonds. But in cells, peptide bonds are never formed between *free* amino acids. Calculate the $\Delta G°$ for the formation of a peptide bond between the free amino acids L-ala and gly.

	ala	+	gly	ala-gly	+	H_2O

$$\Delta G_f^\circ = -88.4 \qquad -88.2 \qquad -117.1 \qquad -56.7$$

Solution:

From Table 5–1, the ΔG_f° for the reactants and products are L-ala, $\Delta G_f^\circ = -88.4$; gly, $\Delta G_f^\circ = -88.2$; ala-gly, $\Delta G_f^\circ = -117.1$; and water, $\Delta G_f^\circ = -56.7$.

$$\Delta G^\circ = (-117.2 + -56.7) - (-88.4 + -88.2)$$

$$= 173.9 - (-176.6)$$

$$= -173.9 + 176.6$$

$$= +2.7 \text{ kcal/mol}$$

Problem 5–20

What is the highest concentration of the dipeptide L-ala-gly that would allow the reaction in Problem 5–19 to occur, assuming that [L-ala] = [gly] = 1.0 mM in the cell?

Solution:

To determine what concentration of the dipeptide product will allow its formation, set $\Delta G = 0$, and let the [dipeptide] = 10^{-x} M.

$$\Delta G = \Delta G^\circ + 1.363 \log \frac{[P]}{[\text{ala}][\text{gly}]}$$

$$0 = +2.7 + 1.363 \log \frac{10^{-x}}{10^{-3} \times 10^{-3}}$$

$$-2.7 = 1.363 \, (-x + 6)$$

$$= -1.363x + 8.18$$

$$1.363x = 10.88$$

$$x = 8.0$$

Therefore, the concentration of the dipeptide in the cell must be less than 10^{-8} M for this reaction to occur between *free* amino acids.

Problem 5–21

Actually, the peptide bond ala-gly is formed in cells using the *activated intermediates* tRNA-ala and tRNA-gly. These tRNA-amino acid compounds contain a high-energy ester bond between the $-COO^-$ of the amino acid and the 2′ or 3′ hydroxyl of the ribose at the 3′-OH end of the tRNA. The formation of each tRNA-amino acid is coupled to the hydrolysis of ATP and release of 7.5 kcal/mol as shown in reactions (1) and (2) below. The peptide bond forms *between the ala and gly residues of the tRNA intermediates* and requires hydrolysis of GTP [reaction (3)], which releases 7.5 kcal/mol.

1. $\Delta G° = -7.5$ kcal/mol
    $$tRNA_1 + ala + ATP \longrightarrow tRNA_1 \, ala + AMP + PP_i$$

2. $\Delta G° = -7.5$ kcal/mol
    $$tRNA_2 + gly + ATP \longrightarrow tRNA_2 \, gly + AMP + PP_i$$

3. $\Delta G° = -7.5$ kcal/mol
    $$tRNA_1 \, ala + GTP + tRNA_2 \, gly \longrightarrow ala\text{-}gly\text{-}tRNA_2 + GDP + P_i + tRNA_1$$

What is the $\Delta G°$ for formation of ala-gly-tRNA$_2$ from ala + tRNA$_1$ and gly + tRNA$_2$?

Solution:

The $\Delta G°$ values for each of the three reactions may be added to find the overall $\Delta G°$ for this coupled reaction.

$$\Delta G° = 3 \, (-7.5) = -22.5 \text{ kcal/mol}$$

Oxidation-Reduction Reactions in Cells Are Coupled Reactions Requiring Cofactors

Oxidation and reduction reactions in the cell are all coupled reactions and provide most of the energy required by cells. *Oxidation* may be defined as the *loss of electrons* and *reduction* as *the gain of electrons.* Generally, electrons move from one molecule to another as part of hydride ions or hydrogen atoms in dehydrogenations in biological systems. The *hydride ions (H:$^-$) or hydrogen atoms (H) are always carried by cofactors instead of being free in the cell.*

H_2 = hydrogen gas (H:H)	H = hydrogen atom (H·)
H^+ = hydrogen ion or proton	H:$^-$ = hydride ion

The movement of electrons, in hydride ions or H atoms, from one compound to another often releases energy that the cell may harness and store. The cofactors that carry the electrons as H$^-$ ions or H atoms are the nicotinamide and riboflavin nucleotides, respectively.

Nicotinamide Cofactors (NAD$^+$, NADP$^+$). Nicotinamide adenine dinucleotide (NAD$^+$) and nicotinamide adenine dinucleotide phosphate (NADP$^+$) are the oxidized forms of these cofactors. Each of these molecules accepts a pair of electrons from a substrate in the form of a hydride ion (H:$^-$) in enzyme-catalyzed dehydrogenation reactions. After accepting a hydride ion, these cofactors are in the reduced form, NAD·H and NADP·H, respectively.

$$Substrate\text{-}H_2 + NAD^+ \longrightarrow Substrate + NAD·H + H^+$$

The hydride ion is generally removed from the substrate as an H atom *plus* an electron, often taken from a carbon atom as shown below in the oxidation of ethanol.

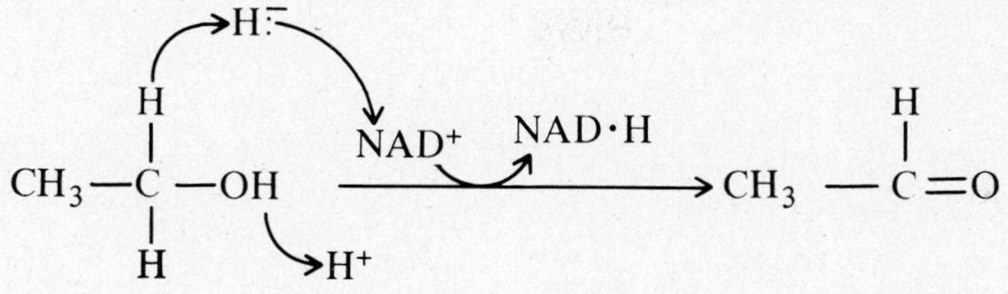

$$CH_3-\underset{\underset{H}{|}}{\overset{\overset{H}{|}}{C}}-OH \quad \xrightarrow[]{NAD^+ \quad NAD\cdot H} \quad CH_3-\overset{\overset{H}{|}}{C}=O$$

with hydride ion $H:^-$ leaving the top and H^+ leaving the bottom.

Problem 5-22

Show the origin of the hydride ion, proton, or H atoms in the following dehydrogenations.

$$\text{lactate} \quad \xrightarrow[\text{LDH}]{NAD^+ \quad NAD\cdot H + H^+} \quad \text{pyruvate}$$

(a)

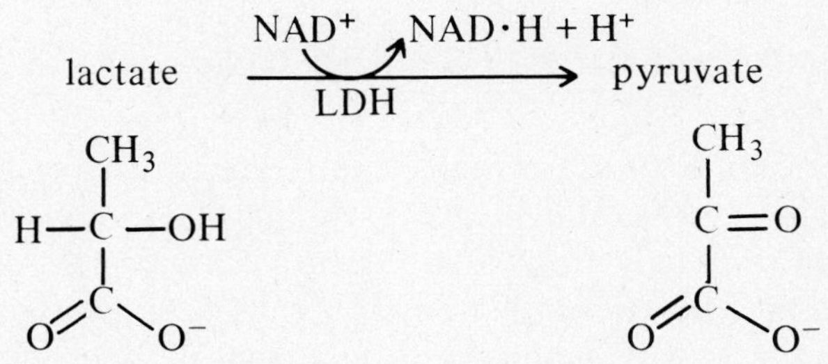

$$H-\underset{\underset{O^-}{\overset{|}{\underset{}{C}}}}{\overset{\overset{CH_3}{|}}{C}}-OH \qquad \qquad \underset{O}{\overset{CH_3}{\underset{\overset{|}{\underset{}{C}}}{C}}}=O$$

(lactate structure with COO⁻) → (pyruvate structure with COO⁻)

(b) $CH_3-(CH_2)_4-COO^- \quad \xrightarrow[\text{FAD} \quad FAD\cdot H_2]{} \quad CH_3-CH_2-CH=CH-CH_2-COO^-$

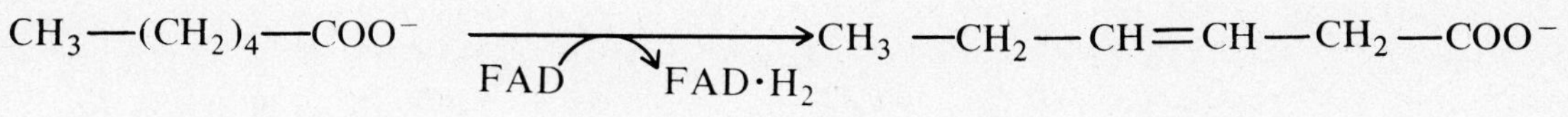

(c) $\quad \text{glucose}-6-P \quad \xrightarrow[NADP^+ \quad NADP\cdot H + H^+]{} \quad \text{phosphogluconate}$

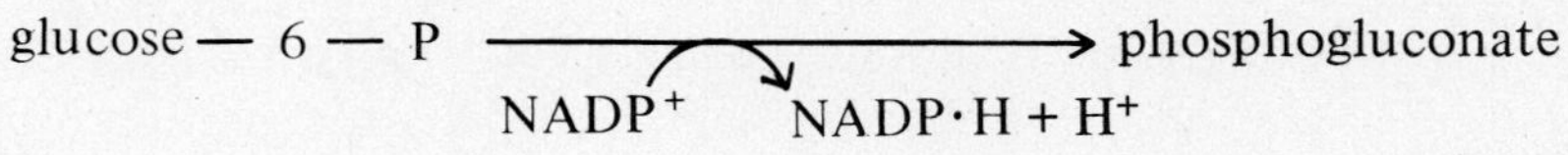

Solution:

(a)

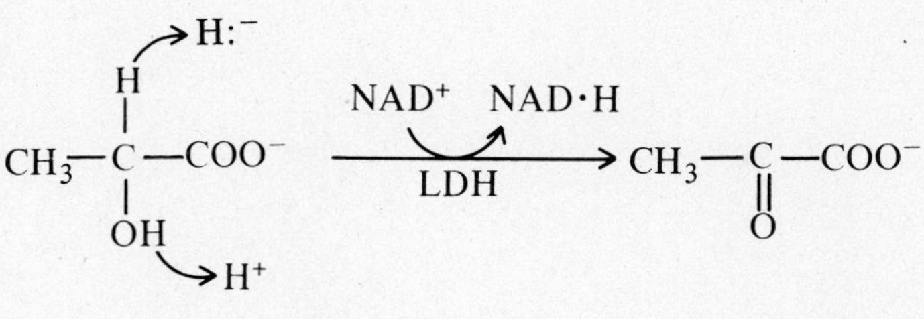

$$CH_3-\underset{\underset{OH}{|}}{\overset{\overset{H}{|}}{C}}-COO^- \quad \xrightarrow[\text{LDH}]{NAD^+ \quad NAD\cdot H} \quad CH_3-\underset{\overset{|}{|}}{\overset{}{C}}-COO^-$$

with $H:^-$ leaving top and H^+ leaving bottom; product is $CH_3-C(=O)-COO^-$

(b)

$$CH_3-CH_2-\underset{\underset{H}{|}}{\overset{\overset{H}{|}}{C}}-\underset{\underset{H}{|}}{\overset{\overset{H}{|}}{C}}-CH_2-COO^- \quad \longrightarrow \quad CH_3-CH_2-CH=CH-CH_2-COO^-$$

$$\xrightarrow[\text{FAD} \quad FADH_2]{}$$

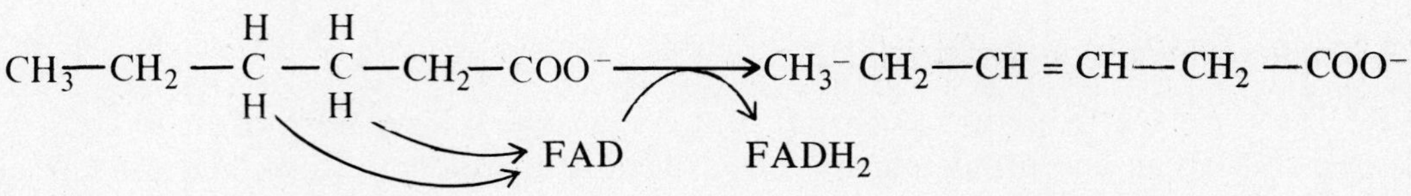

(c)

$$\text{H}_2\text{C}-\text{OP} \qquad \text{H:}^- \qquad \text{H}_2\text{C}-\text{OP}$$

Flavin Cofactors (FMN, FAD). The oxidized flavin cofactors flavin adenine mononucleotide (FMN) and flavin adenine dinucleotide (FAD) accept a pair of electrons from a substrate (H–S–H) as a pair of hydrogen atoms, giving the reduced forms $\text{FMN} \cdot \text{H}_2$ and $\text{FAD} \cdot \text{H}_2$, respectively.

$$\text{Substrate-H}_2 + \text{FMN} \longrightarrow \text{FMN} \cdot \text{H}_2 + \text{Substrate (Oxidized)}$$

Cytochromes. The cytochromes are proteins that contain a heme or a heme derivative as a prosthetic group. The Fe^{2+} in the heme participates in oxidation-reduction reactions allowing electrons to flow singly from reduced cofactors to electron acceptors such as O_2 and NO_4^-. Many of the enzymes in mammalian cells that detoxify foreign chemicals are cytochrome proteins that catalyze the addition of an —OH group to the chemical after reducing O_2.

Problem 5–23

The transfer of electrons from a substrate to NAD^+ or FMN is an example of a coupled reaction. In the following reaction, a pair of electrons (in a hydride ion) are passed from one cofactor to another in an enzyme-catalyzed dehydrogenation. The enzyme is $\text{NAD} \cdot \text{H}$ reductase and is found in the mitochondrial electron transport system of most cells.

$$\text{NAD} \cdot \text{H} + \text{H}^+ + \text{FMN} \longrightarrow \text{NAD}^+ + \text{FMN} \cdot \text{H}_2$$

Show the two separate redox reactions that are coupled in this reaction.

Solution:

Reactions: (1) $\text{NAD} \cdot \text{H} + \text{H}^+ \longrightarrow \text{NAD}^+$

(2) $\text{FMN} \longrightarrow \text{FMN} \cdot \text{H}_2$

Sum: $\text{NAD} \cdot \text{H} + \text{H}^+ + \text{FMN} \longrightarrow \text{NAD}^+ + \text{FMN} \cdot \text{H}_2$

Standard Reduction Potential, E_0

The standard reduction potential of a compound is a measure of its tendency to give up electrons to an acceptor molecule. Hydrogen is the reference molecule and has an arbitrary value of $E_0 = 0.0$. Compounds that tend to give up electrons to hydrogen have $-E_0$ values; molecules that tend to accept electrons from hydrogen have $+E_0$ values. Strong oxidizing agents have $+E_0$, while strong reducing agents have $-E_0$ values.

Note: E_0 values are calculated under standard conditions. These conditions are never found in the cell, but we may use this as a model to estimate what may happen in the cell..

Since oxidation-reduction reactions are coupled reactions, the ΔE_0 between donor and acceptor is used to describe the tendency of electrons to flow from one compound to another in the couple. Table 5–4 contains E_0 values for some common redox reactions in cells.

Table 5–4. Standard Reduction Potentials (E_0) for Some Cellular Redox Reactions

Reaction (Written as Oxidation)	E_0 (V)
α-Ketoglutarate $\longrightarrow$ succinate $+ CO_2 + 2H^+ + 2e^-$	-0.67
Isocitrate $\longrightarrow$ α-ketoglutarate $+ CO_2 + 2H^+ + 2e^-$	-0.38
$NAD \cdot H + H^+ \longrightarrow NAD^+ + 2H^+ + 2e^-$	-0.32
Succinate $\longrightarrow$ fumarate $+ 2H^+ + 2e^-$	-0.13
$FAD \cdot H_2 \longrightarrow FAD + 2H^+ + 2e^-$	-0.05
$H_2 \longrightarrow 2H^+ + 2e^-$	-0.00
Ubiquinone$_{(red)}$ $\longrightarrow$ ubiquinone$_{(ox)}$ $+ 2H^+ + 2e^-$	$+0.05$
Cytochrome $b + Fe^{2+} \longrightarrow Fe^{3+} + e^-$	$+0.10$
Cytochrome $c + Fe^{2+} \longrightarrow Fe^{3+} + e^-$	$+0.28$
Cytochrome $a_3 + Fe^{2+} \longrightarrow Fe^{3+} + e^-$	$+0.40$
[a]$O_2 + 4e^- + 4H^+ \longrightarrow H_2O$	$+0.82$

[a]Written as reduction.

Relationship Between ΔE_0 and $\Delta G°$

Normally, *electrons move toward acceptors with a more positive E_0 value*. The movement of electrons from a compound with a negative E_0 to a compound with a more positive E_0 in a coupled reaction results in energy release. The amount of energy that is released is related to the magnitude of the ΔE_0 between the donor and acceptor reactions. The energy released may be quantitated by relating it to $\Delta G°$ for the coupled reaction and finally to ΔG, the free energy available to do useful work in the cell.

The equation $\Delta G° = -nF \, \Delta E_0$ *allows the energy released by electron flow to be related to the standard free energy change for the coupled reaction.*

$\Delta G°$ = standard free energy change and is determined by subtracting the $\Delta G_f°$ of the reactants (*e* donor) from the $\Delta G_f°$ of the products (*e* acceptor)

n = number of electrons transferred (n = 2 in most biological dehydrogenations; n = 1 in cytochrome Fe oxidation-reductions)

F = Faraday constant, 23,062 calories

$\Delta E_0 = E_0 (\text{acceptor}) - E_0 (\text{donor})$

From the equation

$$\Delta G = \Delta G^\circ + 1.363 \log \frac{[P]}{[R]} \tag{4}$$

the free energy released during the redox reaction may be determined. Since $\Delta G = \Delta G^\circ$ when $[P] = [R]$, then ΔG° is used as an estimate of ΔG.

Problem 5-24

The reaction below is a coupled reaction involving the transfer of a pair of electrons from isocitrate to the cofactor NAD^+ via a hydride ion $H:^-$ and catalyzed in mitochondria by the enzyme isocitrate dehydrogenase.

$$\text{isocitrate} + NAD^+ \xrightarrow{\text{isocitrate DHase}} \alpha\text{-ketoglutarate} + CO_2 + NAD \cdot H + H^+$$

Calculate the ΔG° value for a pair of electrons moving from isocitrate to NAD^+ using the E_0 values in Table 5-4.

Solution:

The ΔG° for this coupled reaction is

$$\Delta G^\circ = -nF \, \Delta E_0 \qquad \text{where } F = 23{,}062 \text{ cal}$$
$$\Delta E_0 = -0.32 - (-0.38) = +0.06$$
$$= -2 \times 23.062 \times 0.06$$
$$= -2.76 \text{ kcal/mol}$$

Problem 5-25

The reaction below is a coupled reaction involving the transfer of a pair of electrons from succinate to the cofactor FAD via two H atoms and catalyzed by the mitochondrial enzyme succinate dehydrogenase.

$$\text{succinate} + \text{FAD} \xrightarrow{\quad \text{succinate DHase} \quad} \text{fumarate} + \text{FAD} \cdot \text{H}_2$$

Calculate the ΔG° value for a pair of electrons moving from succinate to FAD using the E_0 values in Table 5-4 p. 211.

Solution:

The ΔG° for this coupled reaction is

$$\Delta G^\circ = -nF \, \Delta E_0 \quad \text{where } F = 23{,}062 \text{ cal}$$
$$\Delta E_0 = -0.5 - (-0.13) = +0.08$$
$$= -2 \times 23.062 \times 0.08$$
$$= 3.68 \text{ kcal/mol}$$

Problem 5-26

$$\text{NAD} \cdot \text{H} + \text{H}^+ + \text{FMN} \longrightarrow \text{NAD}^+ + \text{FMN} \cdot \text{H}_2$$

Calculate ΔG° for an electron pair moving from NAD·H to FMN. The E_0 values for the reactions NAD·H + H$^+$ → NAD$^+$ and FMN → FMN·H$_2$ from Table 5-4 are -0.32 and -0.05 V, respectively, assuming that FMN and FAD have the same E_0 values.

Solution:

The ΔG° for this coupled reaction is

$$\Delta G^\circ = -nF \, \Delta E_0 \quad \text{where } n = 2$$
$$F = 23{,}062 \text{ cal}$$
$$\Delta E_0 = [-0.05 - (-0.32)] = +0.27$$
$$= -2 \times 23{,}062 \times (+0.27)$$
$$= -12{,}000 \text{ cal/mol}$$
$$= -12.2 \text{ kcal/mol}$$

Problem 5-27

In Problem 5-26, what would the ΔG value be if [products] = [reactants], that is, [NAD·H] [FMN] = [NAD$^+$] [FMN·H$_2$]?

Solution:

Using the equation $\Delta G = \Delta G° + 1.363 \log ([P]/[R])$, since $[P] = [R]$, the ratio $[P]/[R] = 1.0$, the log of $1.0 = 0$, and the term $1.363 \log ([P]/[R])$ becomes zero. Therefore, under these conditions, $\Delta G = \Delta G°$ and $\Delta G = -12.2$ kcal/mol.

Oxidative Phsophorylation: The Electron Transport System and Generation of ATP

The generation of ATP in cells occurs largely by a phosphorylation mechanism that is coupled to the flow of electrons along the electron transport system (ETS). The ETS is a series of molecules bound to the inner mitochondrial membrane. Electrons that are removed during specific oxidation (dehydrogenation) reactions are introduced into the ETS at the nicotinamide or flavin cofactors. These electrons move from NAD or FMN to nonheme iron proteins and on down the chain to reduce O_2.

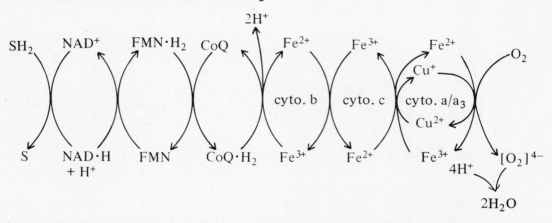

Figure 5–1. Model of the ETS systems found in mammalian mitochondria.

While electrons move as pairs in a hydride ion or two H atoms in the initial part of the ETS, they move singly along the Fe of the cytochromes in the final part. The protons are stripped from the electrons between ubiquinone and cytochrome *b*. The protons are expelled into the space between the mitochondrial membranes, increasing the $[H^+]$ (decreasing pH) in this intramembrane space. The difference between the $[H^+]$ on opposite sides of the inner mitochondrial membrane is called the *proton gradient.* There is potential energy in any gradient and the free energy inherent in the proton gradient is used by ATP synthase to form ATP from ADP and HPO_4^{2-}. Exactly how the energy released by e^- flow is used to form a proton gradient is not understood. Also, the mechanism by which ATP synthase (in the inner mitochondrial membrane) utilizes the energy of the proton gradient has not yet been determined.

The terminal electron acceptor in the chain, O_2, is bound by cytochrome oxidase and receives four electrons, giving transient O_2^{4-}. Four protons are added and two H_2O molecules are released to the mitochondrial matrix. The O_2^- radicals probably remain bound to the cytochrome oxidase since the O_2^- radicals are extremely reactive and will oxidize many cellular molecules if they are free in the cell.

Problem 5–28

Show the atomic structure of an oxygen atom, O_2(gas), O_2^-, and O_2^{4-}.

Solution:

Problem 5–29

Give one reason why the components of the ETS in Fig. 5–1 have been arranged in that order.

Solution:

Many types of experimental data have suggested the arrangement of cofactors in Fig. 5–1. One of the most compelling arguments in its favor is the ascending E_0 values of the components given that the electrons generally flow from negative to positive.

Problem 5–30

In the inner membrane of chloroplasts, an electron transport system has been demonstrated in which a pair of electrons moves from chlorophyll P700 to the Fe-sulfur protein, P430. Calculate ΔG for the movement of this pair of electrons, assuming that the $E°$ for chlorophyll P700 = +0.4 V and the $E°$ for P430 = −0.60 V.

Solution:

$$\Delta E° = -0.60 - (+0.4) = -1.0$$

$$\Delta G° = -nF\,\Delta E° \qquad \text{where } n = 2$$

$$F = 23.0 \text{ kcal/mol}$$

$$= (-2)(23)(-1.0)$$

$$= +46 \text{ kcal/mol}$$

Assuming that the concentration of reactants and products is equal, $\Delta G = \Delta G°$ and $\Delta G = +46$ kcal/mol.

Problem 5–31

In Problem 5–30, suggest a mechanism that would allow the chlorophyll P700 to emit two electrons (which it does) even though ΔG is positive and very large.

Solution:

Energy must be added to the system to allow the electrons to move "against" the energy gradient. In fact, solar energy (light of wavelength 700 nm) absorbed by chlorophyll molecules in chloroplast membranes provides the energy needed to overcome the large positive ΔG.

Problem 5–32

The hydrolysis of ATP to ADP + P_i has an approximate $\Delta G° = -7.5$ to -8.0 kcal/mol. Therefore, $\Delta G°$ for ATP formation is $+7.5$ to $+8.0$ kcal/mol. Between which components of the ETS system shown in Fig. 5–1 would enough energy be released during electron flow to allow the phosphorylation of ADP to ATP?

Solution:

In the equation $\Delta G = \Delta G° + 1.363 \log [P]/[R]$, $\Delta G = \Delta G°$ when [product] = [reactant]. Assuming that [product] = [reactant] in the ETS above, $\Delta G = \Delta G° = -nF \Delta E°$. To determine the steps where $\Delta G > -8$ kcal/mol, we may determine what ΔE_0 value is required to give $\Delta G° > -8$ kcal/mol.

$$-8 = -2 \times 23.0 \times \Delta E_0$$

$$\Delta E_0 = \frac{8}{46} = 0.17$$

Thus, whenever the difference in E_0 is 0.17 or greater, then theoretically, there is enough energy released to drive the phosphorylation of ADP to produce ATP. The ΔE_0 exceeds 0.17 between NAD·H and FMN, cytochromes b and c, and cytochrome oxidase and O_2.

Exercises

The answers are provided at the back of the book.

5-1. Determine ΔG° for the fumarase reaction from:

(a) the G_f° values of the reactants and products in Table 5-1

(b) the $K_{eq} = 4$ for the reaction

$$\text{fumarate} + H_2O \longrightarrow \text{malate}$$

5-2. Calculate ΔG for the reaction catalyzed by lactate dehydrogenase under the following conditions:

$$\text{lactate} + NAD^+ \longrightarrow \text{pyruvate} + NADH + H^+$$
$$100\ \mu M \quad 0.13\ mM \quad\ \ 0.11\ mM \quad 10\ \mu M$$

$\Delta G^\circ = +6.0$ kcal/mol with pyruvate as product.

5-3. The following reaction would proceed in the cell under which of the following conditions (assuming that [B] = [C])?

$$A \longrightarrow B + C \quad \Delta G^\circ = +6.7 \text{ kcal/mol with A as reactant}$$

(a) [A] = 1000 $\times$ [B]

(b) [B] = 1000 $\times$ [A]

(c) $\Delta G < 0$

(d) The K_m for A is less than 1 μM

5-4. Under the following conditions, what concentration of oxaloacetate would be required for the malate dehydrogenase reaction to go toward malate as a product?

$$\text{oxaloacetate} + NADH + H^+ \longrightarrow \text{malate} + NAD^+$$

$\Delta G^\circ = -6.0$ kcal/mol for this reaction with malate as product

$$[\text{malate}] = 40\ \mu M \quad [NAD^+] = 0.15\ mM \quad [NADH] = 90\ \mu M$$

5-5. In experiments to determine the concentration of metabolites in rat livers in different nutritional states, the following data were obtained:

Metabolite	Concentration in Liver (mM)
ADP	1.0
ATP	1.7
3-Phosphoglycerate	0.17

Given these data, what is the highest [1,3-diphosphoglycerate] that will allow the 3-PGA kinase reaction to proceed toward ADP as product in gluconeogenesis? Assume that $\Delta G° = -4.5$ kcal/mol with ATP as product.

1,3-diphosphoglycerate + ADP $\rightleftharpoons$ 3-phosphoglycerate + ATP

5-6. Calculate the ΔE_0 value for the movement of a pair of electrons from cytochrome *a* to O_2.

5-7. Show that there is enough energy released in the reaction in Exercise 5-6 to allow the phosphorylation of ADP to ATP. (Assume that $\Delta G°$ for ATP formation is greater than +7.8 kcal/mol.)

Answers to Exercises

Chapter 1

1-1. (a) 4.410×10^3 (b) 7.384×10^4 (c) 4×10^{-4}

1-2. (a) 1.23 (b) −4.43 (c) 6.3799

1-3. (a) 6.3×10^{-4} (b) 0.32 (c) 199.5 (d) 39.8

1-4. (a) $x = 0.27$ (b) $x = 6.26$ (c) $x = 0.29$ (d) $x = 0.13$ (e) $x = 5.33$

1-5. (a) 0.23 μmol (b) 2.12 μmol

1-6. (a) 3 μmol/mL (b) 0.01 μmol/mL (c) 1×10^{-5} μmol/mL

1-7. 24.08×10^{16}

1-8. (a) 89 mg (b) 1.11 mg (c) 1.33×10^{-3} mg (d) 50 μg (e) 0.04 g

1-9. $A = 1.34 \times 1 \times 0.150 = 0.20$

1-10. (a) 0.66 mg/mL (b) 0.18 mg/mL (c) 0.22 mg/mL

Chapter 2

2-1. $K_a = 10^{-4} \times 10^{-4.3}/[HA] = 1.35 \times 10^{-5}$; [propionic acid] = 0.37 mM

2-2. $K_a = 0.015$

2-3. ratio $[A^-]/[HA] = 85$

2-4. [acetate ion] = 0.067 M

2-5. $[HCO_3^-] = 4.41$ mM

2-6. $[his^0] = 8.6$ mM

2-7. The net charge is zero on most of these peptide molecules at pH 5.1.

2-8. (a) glutamine: pI = 5.65 (b) peptide in Problem 2-7: pI = 5.62

2-9. (a) pH = pK_a for COOH group = 2.34 (b) pH = pK_a for NH_3^+ group = 9.60 (c) pH = pI = 2.34 + 9.60/2 = 5.97

2-10. Na acetate 0.685 g, acetic acid 0.33 g in 500 mL of water.

Chapter 3

3-1. $v_0 = \dfrac{0.6 \times 80}{90 + 80} = 0.28$ μmol/min/mg

3-2. (a) $K_m = 4.5$ μM (b) $V_{max} = 2.6$ μmol/min/mg protein (c) = competitive inhibition (d) $K_i = 5$ μM

3-3. $v_0 = \dfrac{0.075 \times 30}{50 + 30} = 0.028$ μmol/min/mg protein

3-4. (a) $K_m = 4.5$ μM (b) $V_{max} = 2.6$ μmol/min/mg protein (c) = competitive inhibition (d) $K_i = 5$ μM

3-5. (b), (c), (e), (g) are all correct

3-6. 625 μmol

3-7. $v_0 = \dfrac{3 \times 22.8}{25 + 22.8} = 1.43$ μmol/min/0.1 mg enzyme $= 14.3$ μmol/min/mg enzyme

Chapter 4

4-1. (b)

Step	Specific Activity	Fold Purification
Cell extract	2.27	—
Dissolved AS precipitate	3.94	3.94/2.27 = 1.74
After heat treatment	20.5	20.5/2.27 = 9.03
Fractions from Sephadex G-200	121.5	121.5/2.27 = 53.5

(c) The M_r for histidine decarboxylase is larger than 450,000, but the exact M_r can not be determined from these data.

(d) There are two different subunits; one has approximate $M_r = 29,000$. The M_r of other subunit is less than 17,000, but its exact M_r cannot be determined from these data.

4-2. Native phosphoglucoisomerase has an M_r of approximately 130,000 and is composed of two subunits, each with M_r of approximately 65,000.

4-3. Angiotensin I has the following amino acid sequence:

N- asp-arg-val-tyr-ile-his-pro-phe-his-leu-C

4-4. EcoR1 site is arbitrarily set at base pair 1. Hind III site is at bp 29. Bsm I site is at bp 1353. Pst I site is at bp 3609.

4-5. Three isozymes of hexokinase exist, each with M_r between 90,000 and 100,000. Each isozyme appears to contain the same two subunits. One subunit has $M_r = 50,000$ and the other about 45,000. The composition of the three isozymes would probably be (1) 50,000 + 50,000, (2) 50,000 + 45,000, and (3) 45,000 + 45,000.

4-6. The peptide is probably a cyclical structure in which the N terminal forms a peptide bond with the C terminal. Another explanation may be that the N and C terminals have chemical groups attached, such as acetyl or sugar moieties preventing dansyl chloride or carboxypeptidases from reacting with them.

Chapter 5

5-1. (a) $\Delta G^\circ = -200.2 - (-199.5) = -0.8$ kcal/mol (b) $\Delta G^\circ = -1.363 \log K_{eq} = -0.82$ kcal/mol

5-2. $\Delta G = +6.0 + 1.363 \log (10^{-3.96} \times 10^{-5.0}/10^{-4.0} \times 10^{-3.88}) = +4.53$ kcal/mol

5-3. (a) and (c) are correct.

5-4. $\Delta G = -6.0 + 1.363 \log (10^{-4.4} \times 10^{-3.8}/10^{-x} \times 10^{-4.0})$; $x = 8.6$. Then the [OAA] must exceed $10^{-8.6}$ M or 2.5×10^{-9} M.

5-5.   The [1,3-bisPGA] must be less than $10^{-6.8}$ M or 0.16×10^{-6} M for this reaction to proceed toward ADP in gluconeogenesis.

5-6. $\Delta E_0 = E_0$ acceptor $- E_0$ donor $= +0.82 - 0.40 = +0.42$

5-7. $\Delta G° = -nF \Delta E_0 = -2 \times 23 \times 0.42 = -19.32$ kcal/mol

Index

A

Absorbance, 31
Acid, definition, 43
Algebraic equation, 14
Allostery, 137
Amino acids, 67
Amino acid sequencing, 161
Antilogs, 13
Avogador's number, 2

B

Base, definition, 43
Base sequences in DNA, 179
Beer's law, 30
Bronsted acid, 43
Buffers, 59
  common ion effect, 60, 62
  definition, 59
  phosphate, 62, 63
  Tris, 61

C

Competitive inhibitor, 101
Concentrations, 24, 25
Conjugate acid-base pair, 43
Cooperativity, 135
Coupled reactions, 203, 208

D

Dissociation constant, 44

E

Electron transport system (ETS), 214
Electrophoresis, 147
Enzyme
  activity units, 151
  definition, 81
  inhibition, 101-115
  kinetics, Michaelis-Menten, 83-84

  purification, 143-148, 152-156
  specificity, 125-129
  -substrate complex, 82
  units, 151
Equations, 14
Equilibrium constants, 44
Equivalents, 55
Exponential forms, 1
Exponents, 2
E_0, 210

F

Free energy change ΔG, 188

G

Gel exclusion chromatography, 144
Gram, 20

H

Henderson-Hasselbalch
 equation, 48

I

Inhibition constant, K_i, 101
Initial velocity, V_0, 83
Isoelectric point, p^I, 76
Isozymes, 156

K

K_{cat}, 134
K_i, 101
K_m, 83, 84

L

Lineweaver-Burk plot, 90
Logarithms, 6

M

Michaelis constant, K_m, 83, 84

NOTES

NOTES

NOTES

NOTES